GENERAL CHEMISTRY EXPERIMENTS
2nd EDITION

by

Jerry L. Mills

and

Roy E. Mitchell

Texas Tech University
Lubbock, Texas

Morton Publishing Company
925 West Kenyon Ave., Unit 12
Englewood, Colorado 80110

Cover Design and Art Work

by Sheri Davis

Table of Contents

Page

Table of Contents (continued)

INTRODUCTION

This book was written to help you, the chemistry student, learn chemistry while you are working in the chemistry laboratory. Each experiment begins with a brief statement of the objective of the experiment. Background on the experiment will follow next. Here an explanation of the experimental procedure, the chemical reactions, the treatment of data, or other information necessary for a student to understand the experiment and how to perform it well is presented. Then a detailed list of tasks to perform is presented. Finally a sheet to hold the data that you obtain will follow.

In order to learn most efficiently in the laboratory, it is suggested that you get in the habit of doing each of the following:

(1) Read and understand the experiment before you come to the laboratory.

(2) Listen carefully as the instructor explains the experiment at the beginning of the period.

(3) Work efficiently, carefully and accurately as you perform the experiment. Think carefully about the procedures you use. Try to remember properties of all the chemicals that you use; observe all chemical reactions carefully. Be able to write chemical equations for every chemical change that you have observed.

(4) Record all of your data immediately on the data sheet. Have your instructor initial your data sheet before you leave the laboratory.

(5) Clean your desk and the assigned area of the laboratory before you leave.

(6) Complete your laboratory report as quickly as possible. Prepare the report so that it is ready to be turned in when you go to your next laboratory period.

Laboratory Techniques

As you progress from one experiment to the next during this laboratory course, you will learn many new laboratory techniques. In fact learning laboratory skills and techniques is one of the primary objectives of the course. Listed below are some laboratory techniques that are either very important or common that you should be familiar with. Also described is the use of the Spectronic 20 colorimeter.

(1) **Cleaning dishes.** It may seem that chemistry is 90% dishwashing. The figure may be inaccurate, but it is true that dishwashing is very frequent and is necessary to obtain good experimental results. Use soap, tap water, and a brush on most glassware. When the glassware is clean, rinse with tap water to remove the soap. Then rinse with distilled water to remove the chemicals that are dissolved in tap water. Treat distilled water as an expensive reagent; when you waste distilled water, you waste a large amount of energy. Always put the distilled water into a plastic squeeze bottle to dispense it. Use several small portions of water to rinse with; it is more efficient than using large portions. Five 2-ml portions of water will rinse a beaker better than one 10-ml portion or better than two 5-ml portions.

1

(2) **Obtaining reagents.** The reagents, or chemicals, that you will need to perform a particular experiment will be in some central location in the laboratory, usually under the hood. These reagents are for your use as well as the other students in your laboratory and as well as students in laboratory periods before and after yours. It is imperative that these reagents not be contaminated if the experiment that you and others will do is to work. The source of contamination is the careless or thoughtless student who introduces a chemical into the wrong container. Such contamination can occur in many ways; it may be as simple as switching the caps or stoppers for two reagent bottles. If you pour too much of a reagent into a beaker, DO NOT POUR ANY BACK INTO THE REAGENT BOTTLE. The beaker into which you poured the reagent might not be clean, and you could contaminate the entire reagent bottle if you returned any of the chemical back into the reagent bottle. In some instances, a single drop of a foreign chemical in a reagent bottle will prevent an experiment from working properly.

(3) **Inserting glass through a rubber stopper.** In many experiments you will need to insert either glass tubing or a thermometer through a rubber stopper. This procedure, if done incorrectly, is one of the most serious causes of injury in the laboratory, resulting in serious lacerations. The tubing and the hole in the rubber stopper should be moistened with water or glycerol. You should simultaneously twist and push, slowly and carefully, the tubing through the hole. Undue force should not be used. A towel can also be used to help protect your hand.

(4) **Vacuum filtration.** Often gravity filtration is too slow or inefficient. In such cases vacuum filtration is used. The water aspirator provides the vacuum to a Büchner funnel, which is illustrated below.

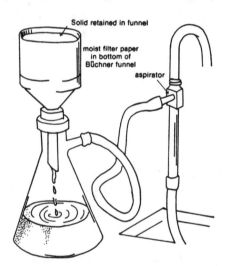

Figure 1. Vacuum filtration

(5) **Pipet.** A pipet is used to measure or deliver an exact volume of solution. The pipet should be cleaned with soap and tap water, and then rinsed with distilled water. When clean, no water droplets should adhere to the inner wall of the pipet. Transfer the solution that you intend to pipet into a beaker. The pipet should not be placed into the reagent bottle. Dry the tip of the pipet, and, using suction from a collapsed rubber bulb, draw several small portions of the solution into the pipet as a rinse. Roll each rinse around in the pipet so the entire inner surface is washed. Discard each rinse solution. To fill the pipet, place the tip well below the surface of the liquid and draw the liquid up into the pipet with the collapsed rubber bulb until the liquid level is 2 to 3 cm above the "mark". Remove the bulb and quickly cover the top of the pipet with your forefinger as in the illustration. Remove the tip of the pipet from the solution and allow the excess solution to drain

2

from the pipet until the level is at the "mark". The bottom of the meniscus should be level with the mark (see below for instructions on how to read the meniscus). Remove any droplet that may be suspended to the outside tip of the pipet. Place the tip of the pipet against the side of the receiving

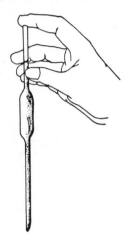

Figure 2. Filled pipete

flask above any liquid level, and allow the solution to drain from the pipet. Do not blow or shake out the last bit of liquid that remains in the tip, as this amount of liquid has been included in the calibration of the pipet.

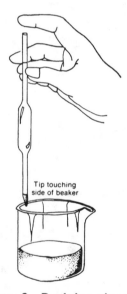

Tip touching
side of beaker

Figure 3. Draining pipete

(6) **Reading the meniscus.** For exact measurements of liquids in pipets, burets, volumetric flasks, or graduated cylinders, the volume should be read from the bottom of the curvature of the liquid, which is referred to as the meniscus. Your eye must be horizontal to the bottom of the meniscus during a reading. The bottom of the meniscus is often more easily read by holding a card behind the liquid level. When recording the level of the bottom of the meniscus on a buret, you should always estimate to tenths between the smallest calibration marks. In the illustration the meniscus is between 6.6 ml and 6.7 ml. By mentally dividing the space between these two marks into ten equal parts, you can estimate that the level of the meniscus is 6.67 ml.

3

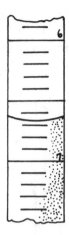

Figure 4. Meniscus in buret

(7) **Balance.** The electric balances that you will use are illustrated below. The instructor will give you exact rules on how to use the balance. If at any time you are not sure how to use the balance properly, ask your instructor. The balances are precision instruments that cost over $1000 each, so be careful with them. Do not drop objects on the pan. Never weigh chemicals of any sort

Figure 5. Top loading balance

directly on the pan; use a beaker, watch glass, weighing paper, or some other container and weigh by difference. If you are weighing something out by difference at one time, it is not necessary to check the zero point of the balance. However, if you are going to weigh something, return to laboratory to preform part of an experiment, and then reweigh to check for a weight loss or gain, then the zero point on the balance must be checked before each weighing. Clean up any spillage of chemicals in the balance area.

(8) **Spectronic 20.** The Spectronic 20 is an instrument used for measuring the effective transmission of monochromatic light through a liquid sample. This measurement determines the concentration of the sample liquid. Two measurements are required: one of a reference liquid (a "blank"), and one of the sample liquid. The ratio of the two measurements is called the "percent transmittance" of the sample. The Spectronic 20 contains a source of white light and an optical system which separates this light into its component wavelengths. Any wavelength may be selected by the operator. The instrument is then "standardized" by filling a sample tube with a reference liquid (such as distilled water), inserting the sample into the instrument, and adjusting the meter to read

4

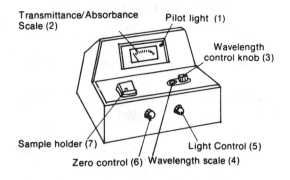

Transmittance/Absorbance Scale (2)
Pilot light (1)
Wavelength control knob (3)
Sample holder (7)
Zero control (6)
Wavelength scale (4)
Light Control (5)

Figure 6. Spectronic 20

"100% transmittance". A sample liquid is then substituted for the reference liquid in the instrument and the percent transmittance is read directly off of the meter.

Operating Procedure

a. Power — Rotate the power switch (6) clockwise. The pilot light (1) will glow. Allow five minutes warm-up time.

b. Wavelength — Adjust the wavelength control (3) to the desired setting as indicated on the scale (4).

c. Zero setting — At this point no light should be reaching the photo-sensitive tube. Therefore there must be no test tube in the instrument and the cover of the sample holder (7) must be closed. Adjust the zero control (6) as required until the meter (2) reads "0" transmittance. The needle should be all of the way to the left.

d. Standardizing light control — Fill a clean sample tube (cuvette) about half-full of pure water or other reference liquid. Insert the cuvette into the sample holder, aligning the mark on the cuvette with the line on the sample holder (7). Close the cover. Adjust the light control (5) as required, until the meter reads "100 %" transmittance. The needle on the meter (2) should be all of the way to the right.

e. Sample measurement — Fill a second clean cuvette about half-full of the sample to be measured. Remove the cuvette of reference liquid and replace it with the cuvette containing the sample. Align the cuvette as above. Close the cover. The Percent Transmittance or Absorbance can now be read directly from the meter (2). Note: it is necessary to repeat step d each time a different wavelength is used. When operating a fixed wavelength, periodically check for meter "drift" from 100%.

5

LABORATORY SAFETY

1. Safety glasses must be worn **at all times** when you are in the laboratory.

2. You should know where to find and how to use first aid equipment. Your instructor will show you where the eye wash and safety shower are located.

3. Consider all chemicals as dangerous unless they are known not to be.

4. If a corrosive chemical gets on your skin or in your eyes, immediately wash the affected area with large quantities of water. Notify the instructor.

5. Never taste anything in the laboratory. Never directly smell the source of a vapor, but rather bring a small quantity of the vapor to your nose with your cupped hand.

6. Reactions involving malodorous, noxious or dangerous chemicals should be performed in the hood.

7. You should know the location of and how to use a fire extinguisher. Small fires, such as a liquid burning in a beaker, can be extinguished by covering the beaker to remove the source of oxygen.

8. No unauthorized experiments are to be performed.

9. Clean up broken glassware immediately.

10. When pouring one liquid into another, do so slowly and cautiously. To dilute an acid, pour **the** acid into the water; *never* pour water into acid.

11. When heating a test tube, make certain that the open end is pointed away from you and your fellow students. Then if overheating causes the contents to bump out, it will not splash anyone.

12. Do not rub your eyes while in the laboratory, as your hands might have chemicals on them.

13. IN CASE OF ANY ACCIDENT, IMMEDIATELY NOTIFY THE INSTRUCTOR. In the event that your instructor is not available, notify the stockroom personnel.

TEST ON SAFE PROCEDURES IN A CHEMISTRY LABORATORY

1. What should be worn in a laboratory at all times to decrease likelihood of eye injury?

2. Only under what conditions are contact lenses permissible in the laboratory? Why?

3. What should you do if something gets in your eyes? What device can you use for help? Where is it?

4. Who should immediately be called for assistance in case of an accident or injury?

5. What should immediately be used if clothing catches fire, or for a large chemical spill on the clothing?

6. How can a small contained fire be put out simply?

7. Describe how to extinguish an open fire in the laboratory. Locate any needed services.

8. Why are unauthorized experiments not permissible?

9. Why must chemicals not be taken from the laboratory?

Test on Safe Procedures, Cont'd.

10. Why is smoking, chewing, eating, or drinking not permissible in the laboratory?

11. Why must reagents be added cautiously?

Name (print)_____

Course No. & Sec. _____

Student Signature _____

Date_____

Experiment 1
Introduction to Chemical Reactions

Object

In this experiment you will perform several chemical reactions and observe the changes in matter which occur during a chemical reaction. Emphasis is placed on observing the substances and the changes in substances which are caused by a chemical reaction.

Background

Chemistry is the science which deals with matter and the changes in matter. In today's experiment you will observe some substances and perform chemical reactions with the substances. The emphasis will be placed on making careful, accurate scientific observations of the materials and the changes in the materials which occur during the chemical reactions. You will be expected to learn something about the properties of the substances you work with today, and also something about how to investigate matter and the changes which occur in matter using scientific procedures. In other words you will learn how to operate like a chemist. Of course you are not expected to operate like a full-fledged chemist yet.

One of the problems you will encounter today (and at many labs in the future) is how to find the sample you are looking for. Your first job today is to obtain a sample of Calcium Carbonate and then you will perform several tests on it. Your job of obtaining this sample will be easier if you know something about Calcium Carbonate. If you know that it is a solid, for example, you can save time by searching only the bottles which contain solids. If you know that the chemical formula for Calcium Carbonate is $CaCO_3$, you know you've got the right bottle when you see either the words "Calcium Carbonate" or the formula "$CaCO_3$" on the label.

Get in the habit of learning the name, the chemical formula, and the state (solid, liquid or gas) of the sample you are looking for. Also try to learn as many properties of the substances you work with as you possibly can. These chemical facts will enable you to get your laboratory work done faster, and will also be of great value to you in your lecture course.

Today you will perform some simple tests on substances. You should perform the experiments carefully but the real emphasis is placed on making careful and complete observations on the changes which occur during the chemical reactions. You will have a data sheet to fill out. Use the data sheet to record all of your observations. Record your observations directly on the data sheet. Get in the habit of putting your data on the sheets and not on scraps of paper, they tend to get lost. At the end of the laboratory period take the data sheet home with you. Study the sheet and complete write up of the experiment, and try to understand everything that happened. Become quite familiar with the procedure and all the substances you deal with in the experiment. This is a desirable practice in all of the experiments which you will perform during the next year in the chemistry laboratory.

Procedure

Carefully record all observations on the experiments to be performed on the data sheet. The detailed directions on the experiment are listed on the data sheet. At this point in the experiment a

9

brief outline of the experiments will be listed. The first experiment deals with calcium carbonate, a solid which has the formula $CaCO_3$. You will obtain a small sample of this solid, place it in a test tube, describe its properties, and then add 2 milliliters of distilled water. Not much will happen at this point in the experiment but you are instructed next to add hydrochloric acid, HCl, which is a liquid. Then an interesting and easily observable reaction should take place. You are then instructed to continue the addition of hydrochloric acid until the change is complete. You are then instructed to write out a description of the chemical reaction which has occurred. You are then instructed to add oxalic acid solution, solid $H_2C_2O_4$ dissolved in water, and observe the chemical reaction which occurs here.

In the next part of the experiment you will perform the same tests using sodium carbonate, Na_2CO_3, as the test material. You will observe this material's behavior upon the addition of water. Then you will observe the reaction upon the addition of hydrochloric acid. You will then see whether the oxalic acid will or will not react with this solution.

The third part of the experiment you will obtain a sample of chalk from the blackboard and repeat the same chemical tests on this material. You will carefully write out all of your observations on the chemical behavior of chalk dust and then write out your conclusions as to whether chalk behaves chemically like calcium carbonate or like sodium carbonate. This is the scientific way to investigate matter.

In the last experiment you will study some colorful solutions. You will investigate a solution containing chromium nitrate, $Cr(NO_3)_3$, dissolved in water. You will observe the color of this solution and then the color changes which occur when you perform several chemical modifications of this material. You will first add sodium hydroxide, which is solid NaOH, dissolved in water. You should obtain a solid from this experiment. You might speculate as to the nature of this solid. You will then add hydrochloric acid and the solid should dissolve. The color of the solution should change and should be different from the solution color you have previously observed. You will then be instructed to perform a test using a new portion of chromium nitrate solution. You will treat this solution with sodium hydroxide and with hydrogen peroxide, H_2O_2, and observe the color change which will occur. You should then observe a color change after the addition of hydrochloric acid and finally another color change upon adding zinc powder to the acidic solution which you have just obtained.

Experiment 2
Scientific Measurements

Object

In today's experiment, you will perform accurate scientific measurements in addition to making scientific observations. You will perform simple measurements of length and mass using both the English and metric systems. You will be asked to perform some calculations on your data. The experiment will acquaint you with the metric system, with methods of recording experimental measurements, and with calculations using significant figures.

Background

You have had some experience in making scientific observations; however, the most useful data which scientists gather is the type obtained by accurate measurements. When workers in different laboratories wish to compare their results, they must use the same system of measurements in expressing the results. You are already familiar with the English system of weights and measures used in this country. All scientists and most other countries in the world use the metric system. This country will eventually drop the English system and adopt the metric system. One of the purposes of today's experiment is to acquaint you with the metric system.

When you are making a scientific measurement, you should utilize all of the accuracy which can be obtained from the instrument you are using. Results should be reported so as to reflect the accuracy of the measurement which you actually made — no more and no less. Results of calculations based on experimental measurements should be done to reflect the accuracy of the results (which depend on the accuracy of the experimental measurements) — no more and no less. In order to do this the simple rules regarding significant figures should be observed when recording data and when calculating with experimental data. In this experiment, the use of significant figures will be emphasized. You should get in the habit of using them correctly in all of your work.

The Metric System

The metric system is based on the meter (length), gram (mass) and liter (volume). Prefixes may be added to indicate multiples of these units. Common prefixes are kilo- (1000 units), and centi- (0.01 units), milli- (0.001 units), and micro- (0.000001 units or 10^{-6} units). The most common unit of length used in laboratory measurements is the millimeter (mm) which is 0.001 meter. Therefore: 1000 mm = 1 m. Meter sticks, 1 m in length are slightly longer than yard sticks, are usually marked in centimeters (cm). 1 cm = 0.01 m = 10 mm. The most common unit of mass is the gram. A kilogram (kg) is 1000 grams and is slightly more than two pounds. For small samples it might be preferred to express the mass in milligrams (mg). A penny has a mass of 3.164 g or 3,164 mg. For volumes, both the liter and milliliter are commonly used in the laboratory. A liter is slightly larger than a quart.

Precision of Measurements and Significant Figures

Most students underestimate the accuracy with which experimental measurements can be made. Most students believe that a buret calibrated with marks at 0.1 ml intervals can be read to ±0.1 ml.

With practice, however, such a buret can be read with an accuracy of ± 0.01 ml (or at least ± 0.02 ml or ± 0.03 ml). All that is required is that when reading the buret the space between 0.1 ml marks be mentally subdivided into 10 parts. The mark closest to the liquid level should be estimated and the volume reading recorded to the nearest hundredth of a milliliter.

In general, before you make any measurement in the laboratory, decide how accurately you can measure with the instrument you are using. Be sure to record all your measurements with this accuracy. If for example, you are reading a buret subdivided to 0.1 ml, record all readings to hundredths of ml. If the liquid level is on the zero mark exactly, record a reading of 0.00 ml to indicate you read it to the nearest hundredth. If you record 0.0 ml, you imply that you only read the volume to the nearest tenth.

When reading most instruments, it will be possible to estimate between smallest subdivisions marked on the instrument. When measuring length with a meter stick subdivided down to mm, report lengths to ± 0.1 mm. When measuring volumes with a graduated cylinder graduated in ml, report volumes to ± 0.1 ml. Balances generally can be operated to read all figures so that estimation of the last figure is unnecessary. The electric top-loading balances used in the laboratories will read to ± 0.001 g, however, it is necessary to check the zero reading in order to obtain this kind of accuracy in each weighing.

Significant figures are the numbers actually read in an experimental measurement. The number 234 ml (which was read in a graduate cylinder to ± 1 ml) contains three significant figures. The number 21.5 ml also contains three significant figures; while 1.7 ml contains only two significant figures. If the three samples above were poured into the same container, how much volume would they occupy? You would calculate the answer to this question by adding the volumes:

$$234 \text{ ml} + 21.5 \text{ ml} + 1.7 \text{ ml} = \text{answer}$$

You can see that the numbers above add to 257.2 ml, but remember that the first sample was only measured to ± 1 ml. The ± 1 ml uncertainty will also be present in the final mixture. To indicate this, we should round off the total volume to 257 ml. In general, the proper procedure in the laboratory is to measure and record your results as accurately as possible, to perform all necessary calculations with your data, and then to express your final answer with the proper number of significant figures.

When adding or subtracting numbers, all numbers must have the same units. You cannot add inches and centimeters. If using exponential notation, all numbers must also have the same exponents. The answer should be written with the same units and the same exponent as all the numbers. The answer should contain the same number of decimals as the least accurate number (smallest number of decimals) in the list. In the above example we add:

$$
\begin{array}{rcl}
234 \text{ ml} & \text{or} & 0.234 \, \ell \\
21.5 \text{ ml} & & 0.0215 \, \ell \\
\underline{1.7 \text{ ml}} & & \underline{0.0017 \, \ell} \\
257.2 \text{ ml} & & 0.2572 \, \ell
\end{array}
$$

rounded to 257 ml   rounded to 0.257 ℓ

When expressed in ml, the least accurate volume was 234 ml which contained no decimals. When these volumes were expressed in ℓ, the 0.234 ℓ volume contains three decimals and the answer should contain only three decimals.

Some additional examples illustrate the procedure.

$$
\begin{array}{lll}
\begin{array}{l}10.1 \quad \text{mm} \\ .105\ \text{m} \\ \underline{1.284\ \text{cm}} \\ \text{cannot be} \\ \text{added}\end{array} \text{or}
&
\begin{array}{l}10.1 \quad \text{mm} \\ 105 \quad\ \text{mm} \\ \underline{12.84\ \text{mm}} \\ 127.94\ \text{mm} \\ \text{rounded off to} \\ 128\ \text{mm}\end{array} \text{or}
&
\begin{array}{l}.0101\ \text{m} \\ .105 \quad\ \text{m} \\ \underline{.01284\ \text{m}} \\ .12794\ \text{m} \\ \text{rounded off to} \\ .128\ \text{m}\end{array}
\end{array}
$$

$$
\begin{array}{llll}
\begin{array}{l}231 \quad\ \text{ml} \\ 50.1\ \ \text{ml} \\ \underline{6.02 \times 10^{-3}\ \ell} \\ \text{cannot be} \\ \text{added}\end{array} \text{or}
&
\begin{array}{l}231 \quad\ \text{ml} \\ 50.1\ \ \text{ml} \\ \underline{6.02\ \text{ml}} \\ 287.12\ \text{ml} \\ \text{rounded off to} \\ 287\ \text{ml}\end{array} \text{or}
&
\begin{array}{l}0.231\ \ell \\ 0.0501\ \ell \\ \underline{0.00602\ \ell} \\ 0.28712\ \ell \\ \text{rounded off to} \\ 0.287\ \ell\end{array} \text{or}
&
\begin{array}{l}231 \quad\ \times 10^{-3}\ \ell \\ 50.1\ \ \times 10^{-3}\ \ell \\ \underline{6.02 \times 10^{-3}\ \ell} \\ 287.12 \times 10^{-3}\ \ell \\ \text{rounded off to} \\ 287 \times 10^{-3}\ \ell\end{array}
\end{array}
$$

When multiplying or dividing numbers different rules should be applied. Units on the answer must be worked out each time. Sometimes units will cancel; at other times fairly complex units will appear in the answer. For example:

$$(0.825\ g/ml)(12.5\ ml) = 10.3\ g$$

$$\frac{(22.4\ \ell)(1.00\ atm)}{(1.00\ mole)(273.15\ deg)} = 0.0820\ \frac{\ell\ atm}{mole\ deg}$$

$$52.4\ ml\ soln \times \frac{40.0\ g\ NaOH}{1000.\ ml\ soln} \times \frac{60.0\ g\ acetic\ acid}{40.0\ g\ NaOH} = 3.14\ g\ acetic\ acid$$

To explain the rules, we need to discuss significant figures further. The following contain three significant figures: 0.0825 g/ml, 12.5 ml, 10.3 g, 1.00 atm, 0.0820 ℓ atm/mole deg and 40.0 g NaOH. The number 1000. ml soln contains four significant figures and the number 273.15 deg contains five significant figures. Do *not* count decimals. The number 1000. ml contains four significant figures whether it is written as 1000. ml or 1.000 ℓ. Zero's used to place the decimal point are not significant figures; for example, expressing a mass of 54.3 mg which can also be written as 0.0543 g shows three significant figures in the measurement.

In the above example, a measurement of 54.3 mg or 0.0543 g with an uncertainty of ± 0.1 mg or ± .0001 g is indicated. This measurement has an uncertainty of 1 part in 543 parts (.1 mg in 54.3 mg) or about 0.2%. When this measurement is combined with other numbers in multiplication or division the answer can be no more accurate than 0.2%. To reflect the accuracy of a multiplication or division answer including this number, only three significant figures should be used in the answer.

In general, any answer for multiplication or division problems should be written with the same number of significant figures as the least accurate number (smallest number of significant figures) used in the problem. Thus, the answer will reflect the accuracy of the actual measurements used in the experiment.

To summarize, when data is added or subtracted, convert all numbers to the same units, perform the arithmetic, and round off to have an answer with the smallest number of decimals in any of the

17

data. When data is multiplied or divided, perform the arithmetic, and round off to have an answer with the smallest number of significant figures in any of the data.

Experimental Procedure

In this part of the write up of the experiment, you will usually find a detailed list of the tasks to be done in the laboratory. In this particular experiment, however, this detailed list is on the data sheet. This part of the write up will be used to describe and explain the tasks you will perform in the laboratory.

Part 1 involves measurements of linear dimensions of your lab manual. You will use both English and metric units. The linear dimensions will be used to calculate the volume of the lab manual. You will also use your measurements to obtain conversion factors for centimeters into inches and liters into gallons. You will find it interesting to compare your experimental values to the accepted ones. You should also find the metric unit calculations easier than the English unit calculations.

Part 2 involves measurements on an ordinary sugar cube. You will measure the linear dimensions and the weight of the cube. You will then calculate the volume, and density of a single cube, and the number of cubes in one kilogram and in one pound of sugar. You may then measure the volume change which results when the cube is dissolved in water, and answer several questions about the solution which results. You will probably be unable to completely explain the volume change.

Part 3 involves measurements on small Pyrex glass beads. You will first attempt to measure the diameter directly. You will measure values for six beads, and use the average value to calculate the volume of a single bead. Later you will measure the volume of fifty glass beads by measuring the volume of water displaced. You will then have to judge these two methods for measuring the volume of a bead. Your answer should be justified in great detail.

You will also weigh fifty glass beads, and calculate the weight of a single glass bead. You will also calculate the density of a single glass bead. You may find it interesting to compare your value to the density of Pyrex glass which is 2.23 g/ml.

By measuring the apparent volume of the glass beads (the actual volume of the glass beads plus the void volume between the beads) during the experiment, you will also gather some data of use in the study of solids. Some solids may be considered to be solid spheres packed as tightly as possible. Even though the spheres are touching, there will be void spaces. Your data on the per cent void volume may shed some light on the efficiency with which hard spheres are packed under less than ideal conditions.

Part 4 is a detailed study of a candle. You will observe the candle before combustion, and during combustion. You will observe chemical changes — ones in which molecules are changed. You should also observe physical changes — ones in which molecules are not changed. You should obtain evidence for two products of combustion. You will write out a balanced chemical equation to show the combustion process.

When you extinguish the candle and remeasure it, you may find some properties which change a lot, and some properties which don't change much. You may be thinking of extensive properties and intensive properties at this point. Extensive properties are ones which depend on the size of the sample (mass and volume, for example). Intensive properties are ones which do not depend on the sample size (density, for example). Which type of property would be the same for all candles? Which type of property would be unique for the candle you use in the experiment? Which type of property would be the same for fifty and for one glass beads?

Name_____   Section_____

Station_____   Date_____

Scientific Measurements

Part 1. A. Remove this data sheet from the lab manual. Close the lab manual and measure the length, width, and thickness of the book using the inch scale and the centimeter scale. Be sure to read and record the measurements using the best accuracy possible from the equipment you use.

length,_____inches;_____ cm

width,_____inches:_____cm

thickness,_____inches;_____cm.

B.   Calculate the following from your data:

The volume of the book in cubic inches. Show your calculations below.

The volume of the book in cubic centimeters. Show your calculations below.

The volume of the book in gallons. Show your work.

The volume of the book in liters. Show your work.

The number of cm in one inch. Use your most accurate measurements. Show your work.

The number of liters in one gallon. Show your work.

How many significant figures in each of the following:

22.4 liters,_____; 2.24 x 10^4 ml,_____; 24.8cm,_____;

25 5/16 inches,_____ ; 2 pounds 5 oz.,_____; 3 qts 7 ozs,_____.

19

Part 2. A. Obtain a sugar cube and measure the length, width, and height of the cube in centimeters. Measure the weight in grams.

length,_____ cm; width,_____ cm; height,_____ cm; weight,_____ g.

B. Calculate the following from your data:

The volume of one sugar cube in ml. Show your work below.

The density of a sugar cube in g/ml. Show your work below.

The number of cubes in one pound of sugar. Remember that one pound is 454 g.

The number of cubes in one kg of sugar.

C. Obtain a 50 ml graduated cylinder and put about 25 ml of water in it. Read the volume exactly. Place the cube in the water and observe the change which occurs. Stir the mixture and read the volume again.

initial volume reading,_____ ml; final volume reading,_____ ml.

Observations upon placing the cube in the water:

Volume change is_____ ml.
Explanation of the change which occurred:

D. Density of the final solution is_____ g/ml. State your assumptions and show your work below.

The final solution is_____% sugar. Show your work below.

Name_____   Section_____

Part 3. A. Obtain about sixty small Pyrex glass beads. Measure the diameter of six beads chosen at random.

_____,_____,_____,_____,_____,_____.

Calculate the average diameter and average radius from the data.

average diameter,_____cm; average radius,_____cm.

Calculate the volume of a single glass bead. Recall that $V=\frac{4}{3}\pi R^3$ for a sphere.

B. Obtain the weight of fifty glass beads; calculate the weight of one glass bead.

Fifty beads weigh_____g; one bead weighs_____g.

C. Place about 25 ml of water in a clean, 50 ml graduated cylinder. Record the volume of the water exactly. Add fifty glass beads carefully (without splashing water out). Record the final volume of the mixture.

initial volume,_____ml; final volume,_____.

The actual volume of fifty beads is_____ml; the actual volume of one bead is_____ml.

Which of the two values for the volume of one glass bead is more accurate?_____
Explain why you believe this.

What would happen if you repeated this experiment but metal beads of the same diameter and four times greater density and mass were used in place of the glass beads?

D. The apparent volume of fifty beads is_____ml. This volume corresponds to_____ml

of actual beads, and_____ml of void volume. The void volume is_____% of the

apparent volume. Show your work below.

When you have finished, dry the beads and return them to a labeled container.

21

Part 4. A. Obtain a candle. Write out a general description of the candle.

Measure the weight of the candle._____g

Measure the volume of the candle._____ml. Explain the method you used. Justify it as the best method to use. Record the measurements and show the calculations in the space below.

B. Light the candle. Describe the flame.

Describe the shape of the solid wax and the molten wax under the flame. Explain these shapes.

Is the melting of wax a chemical change or a physical change? Write out two observations which support your answer.

Light a match and keep it ready. Blow out the candle flame and then quickly bring the match flame close to (but not touching) the candle wick. Write out your observations below. What evidence do you now have for paraffin wax in the gaseosus state.

C. Use your crucible tongs to hold a cold crucible about 4 cm above the flame. What forms on the surface of the crucible?

Suspend a drop of barium hydroxide, $Ba(OH)_2$, solution near the flame using a nichrome wire loop. Observe the drop closely. The formation of barium carbonate, $BaCO_3$, as a white film on the drop is due to a reaction with carbon dioxide, CO_2. Does the candle flame give off carbon dioxide? Does your breath give off carbon dioxide?

What is actually burning in a candle?

Write a balanced chemical equation for the combustion process. You may assume paraffin wax is $C_{50}H_{102}$, $C_{49}H_{100}$, or $C_{48}H_{98}$.

D. Blow out the candle. Let it cool and reweigh. _____g

Measure the volume of the candle as before. Show your data below.

Calculate the density of the candle before combustion. _____g/ml

Calculate the density of the candle after combustion; _____g/ml

Tabulate the changes in the candle which occurred during the combustion:

weight,_____

volume,_____

density,_____

other,_____

Prepare a list of properties which changed a lot on the left, and properties which changed very little on the right. Even if you did not measure them, list the following properties: mass, length, diameter, color, shape, density, volume.

properties which changed a lot   properties which changed very little

List properties which would be different for a sample containing fifty beads and a sample which contains only one bead.

List properties which would be the same for these samples.

List some of the intensive properties you found for a glass bead.

8. Observations upon heating SiO_2 sample.

Weight of first dish + dry SiO_2, _____ g

Weight of SiO_2, _____ g = _____ g - _____ g

%SiO_2 in sample = _____ %. Show your calculations below.

9. Describe the solution (NaCl, NH_4Cl, water) in the second dish.

Describe the changes by evaporating water from the mixture.

Weight of second dish + residue (NaCl, NH_4Cl), _____ g

10. Describe the changes by subliming NH_4Cl from the residue.

Weight of second dish + final residue (NaCl), _____ g

Weight of NaCl, _____ g = _____ g - _____ g

% NaCl = _____ %. Show your calculations below.

Weight of NH_4Cl, _____ g = _____ g - _____ g

% NH_4Cl = _____ %. Show your calculations below.

11. Summary of Results of Method B.

Weight of sample = _____ g

Weight of SiO_2 = _____ g, % SiO_2 = _____%

Weight of NaCl = _____ g, % NaCl = _____%

Weight of NH_4Cl = _____g, % NH_4Cl = _____%

Sum of weights = _____g, Sum of % = _____%

Deviation between sample weight and sum of weights = _____g.

How do you account for this deviation?

12. Comparison of Method A and Method B.

(a) Ease of separating the components.

(b) Accuracy of results.

(c) Errors in the methods.

Overall comparison of the methods.

Experiment 4
Paper Chromatography

Object

Chromatography, a powerful technique used to make separations, will be used to isolate each of three metal cations that are originally present as a mixture.

Background

One of the most common of all problems that arises in a chemistry-related laboratory is the separation of one substance from another or a mixture of others. While there are many techniques that can be used, few are more powerful than chromatography. Chromatography, then, is a separation technique, and can be generally defined as a method of separating components of a solution by distributing them between two immiscible phases. The two immiscible phases can be, for example, a stationary solid phase such as paper with a moving liquid phase. If the components of a mixture differ in their attraction to the solid stationary phase (paper) but have about the same attraction for the liquid phase, then the components of the mixture will be selectively picked up by the moving phase as it passes over the stationary phase, thus effecting separation. In most general terms the moving phase can be a gas or a liquid and the stationary phase can be a solid or even a liquid held stationary by some solid to which the liquid is strongly absorbed.

In this experiment the moving phase will be a liquid that is a mixture of acetone, water, and hydrochloric acid. The solid phase will be a high quality filter paper. The liquid phase will travel along the paper by capillary action. A mixture of three metal cations (copper, iron, and nickel) which are initially introduced onto one spot on the paper will move along the paper at different speeds (because each is attracted to the wet paper to a different degree). What was originally a mixture can thus be separated into three ions. This is graphically illustrated in Figure 4-I.

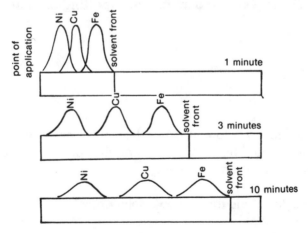

Figure 4-I. Movement of compounds down the paper

Qualitatively the components can be identified by comparing the unknown chromatograms (the paper with the separated ions) with chromatograms made from known components suspected to be in the unknown mixture.

33

The three ions in this experiment are copper (Cu^{2+}), nickel (Ni^{2+}) and iron (Fe^{3+}). Cu^{2+} ions in water react with ammonia (NH_3) to form a deep blue-colored complex $Cu(NH_3)_4^{2+}$. Thus wet paper containing Cu^{2+} ions will turn a deep blue when exposed to gaseous NH_3. The deep blue color can be used to indicate the presence of Cu^{2+} ions and their position on the paper. Ni^{2+} ions react with a reagent called dimethylglyoxime to produce a red color. This colored complex can be used to identify the presence and location of Ni^{2+} ions on the paper. Fe^{3+} ions on wet paper produce a rust color, so that no color reagent is needed to develop a color for these ions.

Cu^{2+} - Blue
Ni^{2+} - Red
Fe^{3+} - Rust

Experimental Procedure

1. Obtain an unknown from your instructor and run it simultaneously with the experiment containing a mixture of all three ions. The unknown may contain one, two, or three of the possible ions Cu^{2+}, Ni^{2+}, or Fe^{3+}.

2. Cut a wick 1 cm wide in the center of the filter paper provided (Figure 4-II). Apply 1 drop of each of the three know solutions Cu^{2+}, Ni^{2+}, and Fe^{3+} to the center of the filter paper. Each of the three

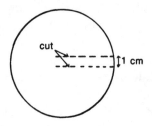

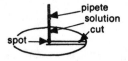

Figure 4-II. Preparation and spotting of paper

spots should be allowed to dry before placing a drop of the next ion directly on top of the previous one. Considerable care should be used in applying the spots, since poor separations result from diffuse or waterlogged samples on the paper.

3. Note and record the color of the three known solutions.

4. Pour the solvent mixture into an evaporating dish about one quarter full. Rest the treated filter paper on the edges of the evaporating dish with 1 cm wick bent down into the solvent. Cover the dish with another dish (Figure 4-III). It is important that the liquid move uniformly through the paper. Allow the experiment to stand undisturbed until the solvent front has moved to the edge of the dish (15-30 min). Discard the solvent at the end of the experiment.

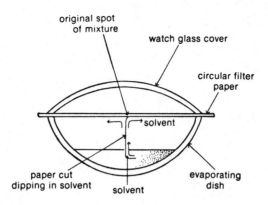

original spot of mixture

watch glass cover

circular filter paper

solvent

paper cut dipping in solvent

solvent

evaporating dish

Figure 4-III. Set up for paper chromatography

5. Remove the paper and allow the solvent to completely evaporate. No color development is needed to identify one of the ions. Which one? Mark the average value for this ring front with a ball pen, averaging out the irregularities.

6. In order to identify the second ion, pour, *under the hood*, a few ml of concentrated ammonium hydroxide into an empty dish top and rest the paper over it, with the wick kept *out* of the solution. What color is observed as a result of the ammonia gas. What species is responsible for this color? Mark the ring front.

7. The coloring agent dimethylglyoxime can be used to identify the third ion. Dip a fresh piece of filter paper in a 1 percent dimethylglyoxime solution and, using it as a brush, streak it across the test paper from center out. Why can dimethylglyoxime be termed a specific reagent in this experiment? Mark the ring front where the color appears. Show your filter paper chromatogram to your instructor.

8. Repeat the experiment with your unknown, following steps 4-7 after you have placed a spot of your unknown in the center of a fresh piece of prepared filter paper.

Do experiment together

35

Before   Unknown #2

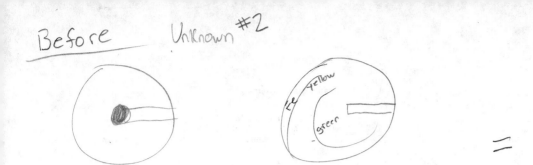

Fe Yellow

green

=

Experiment 5
Formula of Magnesium Oxide

Object

The formula of magnesium oxide will be determined by measuring the weight of oxygen which combines with magnesium, and combining this information with the known atomic weights of magnesium and oxygen.

Background

Even though the Law of Definite Composition and the Law of Multiple Proportions were well established laws in chemistry in 1808, John Dalton's "elegant and probable theory" — the atomic theory — was not accepted by some chemists for many years. We now know that some of the problem lay in these workers inability to distinguish between weights of elements known in modern language as:

equivalent weight — the weight of an element which will combine with eight grams of oxygen

atomic weight — the weight of an element based on a scale of the most common isotope of carbon equals twelve

molecular weight — the weight of a molecule of an element.

	equivalent weight	atomic weight	molecular weight
hydrogen	1	1	2 (H_2)
oxygen	8	16	32 (O_2)
nitrogen	(varies)	14	28 (N_2)
potassium	39	39	39 (K)
sulfur	16	32	256 (S_8)

The equivalent weight of some elements can vary if more than one compound can be formed, e.g., nitrogen in ammonia (NH_3) and nitrogen (III) oxide (N_2O_3) shows an equivalent weight of 4.67 (or 14/3), while in nitrogen dioxide (nitrogen (IV) oxide, NO_2) nitrogen shows an equivalent weight of 3.5 (or 14/4).

In this experiment you will measure the weights of magnesium and oxygen which combine. From the known atomic weights, you will calculate the formula for magnesium oxide. Performing this experiment will also acquaint you with some magnesium chemistry. You should observe a reaction between magnesium and nitrogen, even though nitrogen rarely reacts with other substances.

Experimental Procedure

1. Clean and dry a crucible and cover.

2. Select a balance and be sure to use the same balance throughout this experiment. Be sure to check the zero point prior to each weighing. See your instructor if your balance does not zero. Weigh the crucible and cover to ± 0.001 g.

3. Obtain a magnesium sample of about 0.5g. Wind the magnesium into a tight spiral which fits the bottom of the crucible.

4. Weigh the crucible and cover and magnesium as before.

5. Place the crucible on a pipe-stem triangle on a ring stand. Place the cover on the crucible to leave a small gap. Heat the crucible gently with a Bunsen burner for about five minutes.

6. Heat the crucible as strongly as possible for ten minutes. **CAUTION** — Magnesium can emit high intensity light when it reacts with air. To avoid eye damage, do not look into the crucible during the reaction.

7. Cool the crucible. Reweigh the crucible and cover and contents as before.

8. Add a few drops of distilled water to the contents of the crucible. Carefully smell the product of this reaction. (A wise procedure to use in smelling anything in the laboratory: pull a handful of the gas to be smelled to the nose — do not stick your nose in the gas to be smelled until you are sure that it doesn't smell too bad.)

9. Heat the contents of the crucible gently for about five minutes and strongly for about five minutes.

10. Cool the crucible and reweigh as before.

11. Reheat the crucible, cool and reweigh. If the weight is not constant to ± 0.008 g; repeat until successive weights are constant within ± 0.008 g.

12. Repeat the procedure using a second magnesium sample.

Name_____ Lab Section_____

Station No. _____ Date_____ _____

The Formula of Magnesium Dioxide

Observations	First trial	Second trial
Weight of crucible and cover	_____	_____
Weight of crucible and cover and Mg	_____	_____
Weight of crucible and cover and contents (first heating)	_____	_____
Weight of crucible and cover and contents (second heating)	_____	_____
Weight of crucible and cover and contents (third heating)	_____	_____
Weight of crucible and cover and contents (fourth heating)	_____	_____

Appearance of Mg: _____

_____

Appearance of contents after first heating: _____

_____

Odor of gas emitted when water added:_____

_____

Appearance of contents after second heating: _____

_____

RESULTS (show your work)	First trial	Second trial
Weight of Mg in compound:	_____	_____
Moles of Mg in compound:	_____	_____
Weight of 0 in compound:	_____	_____
Moles of 0 in compound:	_____	_____
Moles Mg per mole of 0 in compound:	_____	_____

41

Correct formula for magnesium oxide _____.

Balanced Equations for the following processes:

Mg + Oxygen

Mg + Nitrogen

Magnesium oxide + water:

Magnesium nitride + water:

Heating wet substances:

Experiment 6
Determination of the Formula of a Hydrate

Object

In this experiment the mole ratio of water to barium chloride will be determined for the hydrate $BaCl_2 \cdot XH_2O$.

Background

Many compounds will dissolve in water, and when the water is evaporated, a solid remains. Some of the compounds, even though they are solids, contain a certain number of water molecules weakly bound within the crystal. The molecules are called *hydrates*, and the number of water molecules "held" by a particular compound is usually a constant. The number of water molecules held does vary from one compound to another, however. For example the formula $CuSO_4 \cdot 5H_2O$ means that there are exactly five water molecules for every one molecule of $CuSO_4$, or stated another way, there are exactly five moles of water molecules for every mole of copper sulfate molecules. Stated still another way, there are 90 g of water (since five moles of water molecules weighs 90 g) for every 160 g of $CuSO_4$ (since one mole of copper sulfate weights 160 g). Water molecules of hydration can often be easily driven off by heating. Thus you would expect that if you weighed out 250 g of $CuSO_4 \cdot 5H_2O$ (the weight of one mole of $CuSO_4 \cdot 5H_2O$ molecules), and heated it until all of the water was driven off, the weight of the remaining $CuSO_4$ would be 160 g.

$$\underset{\substack{\text{Mol. Wt. 250}}}{\overset{\substack{250 \text{ g}}}{CuSO_4 \cdot 5H_2O}} \xrightarrow{\Delta} \underset{\substack{\text{Mol. Wt. 160}}}{\overset{\substack{160 \text{ g}}}{CuSO_4}} + \underset{\substack{\text{Mol. Wt. 18}}}{\overset{\substack{90 \text{ g}}}{5H_2O}}$$

The number of moles of water of hydration vary from one compound to another. Examples include $NiSO_4 \cdot 7H_2O$, $NaBr \cdot 2H_2O$, $Na_2CO_3 \cdot 10H_2O$, and $CoCl_2 \cdot 6H_2O$. In this experiment you are going to determine the number of moles of water that are combined with each mole of barium chloride in the hydrate $BaCl_2 \cdot XH_2O$, that is, the value of X. As in the example above, when a hydrate is heated, water is lost.

$$BaCl_2 \cdot XH_2O \xrightarrow{\Delta} BaCl_2 + XH_2O$$

If the weight of water lost by heating a known weight of $BaCl_2 \cdot XH_2O$ can be determined, then it is a simple calculation to determine the value of X. The weight of the water is the difference between the weight of the barium chloride before and after heating.

$$\text{\# of moles of water} = \frac{\text{wt. of } H_2O}{\text{mol. wt. of } H_2O} = \frac{(\text{wt. of } BaCl_2 \cdot XH_2O) - (\text{wt. of } BaCl_2)}{18 \text{ g/mole}}$$

$$\text{Value of X} = \text{mole ratio of water to } BaCl_2 = \frac{\text{\# of moles of water}}{\text{\# of moles of } BaCl_2}$$

$$\text{\# of moles of } BaCl_2 = \frac{\text{wt. of } BaCl_2}{\text{mol. wt. of } BaCl_2} = \frac{\text{wt. of } BaCl_2}{208 \text{ g/mole}}$$

Experimental Procedure

1. Obtain a crucible and cover. Clean and dry them. For the rest of the experiment, the crucible and cover should be handled only with crucible tongs, so as to prevent moisture from your hands from changing their weight. Heat the crucible and cover with a burner for 1 min to drive off any traces of water. Let them cool to room temperature. (A hot crucible causes convection currents that will yield erroneous weights.)

2. Weigh the crucible and cover to 0.001 g. Record the weight.

3. Add 2 to 3 grams of $BaCl_2 \cdot XH_2O$ in the crucible, replace the cover, and weigh the crucible, cover, and contents to the nearest 0.001 g. Record the weight.

4. Put the crucible on a triangle on a ring stand, leaving the cover open so that water can escape and so that you can observe what is happening. (Figure 6-I) Heat the crucible very gently with the

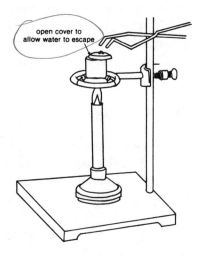

Figure 6-I. Apparatus to decompose hydrate

44

burner for about 5 min. Heat more strongly for about another 5 min. Finally put the burner directly under the crucible and heat for 10 min with the blue cone of the flame (the hottest part).

5. Remove the crucible and cover from the flame. Cover the crucible with its cover, and allow them to cool *completely* to room temperature (at least 10 min). Weigh the crucible, contents and cover, recording the weight.

6. In order to insure that a constant weight has been obtained, put the crucible and cover back on the triangle and heat for another 5 min. Repeat the cooling and weighing, step 5. Record the weight. This weight should be very close to the weight observed after the first heating treatment. Using the data obtained, calculate the number of moles of water per mole of barium chloride in your sample.

Experiment 7

Gas Laws

Object

In this experiment, you will measure the volume and temperature of a gas at constant pressure (Charles's Law) and the volume and pressure of a gas at constant temperature (Boyle's Law).

Background

In the first part of this experiment, you will assemble the apparatus shown in Figure 7-I. The 50 ml erlenmeyer flask with 10 ml syringe attached will hold an air sample at a constant pressure. When

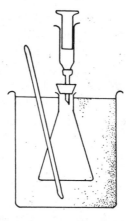

Figure 7-I. Charles's Law set up

the syringe is near the zero mark, about 50 ml (total volume) of air is in the apparatus. When the syringe is near the 10 ml mark about 60 ml of air is in the apparatus (50 ml in the flask and 10 ml in the syringe). The apparatus is set up to measure the volume of the gas sample at various temperatures.

The volume-temperature relationship for a gas sample was described by Jacques Alexandre Cesar Charles and is usually called Charles' Law. Your data for total volume of the gas, V, and temperature will be graphed. You will use Celsius temperature, t, for the x-axis. When dealing with gases, however, the Absolute or Kelvin scale must be used. Absolute temperature, T, is related to Celsius temperature, t, by the relationship:

$$T = t + z$$

You will determine a value of z from your data and compare it to the accepted value of 273.15°. Charles' Law can be stated mathematically as $V \alpha T$ or $V = kT = k(t + z)$. You will determine a value of z from the graph of V vs t. Draw a straight line through the data points and extend it to $V = O$. At this point, volume and absolute temperature are equal to zero. The sum of t (a negative temperature on the Celsius scale) and z is zero. The value of z is the negative of the Celsius reading at $V = O$.

In the second part of the experiment, you will set up the apparatus shown in Figure 7-II. An air sample in the 30 ml syringe is held at constant temperature while the volume and pressure are varied.

49

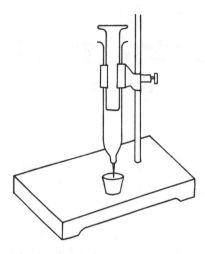

Figure 7-II. Boyle's Law set up

Your first reading will be the volume of air when the pressure is equal to atmospheric pressure. This corresponds to a relative pressure of zero, but the absolute pressure is actually equal to barometric pressure. Your next reading will be obtained by balancing one book on the syringe plunger. The relative pressure and the absolute pressure will be "one book" higher than the previous pressure. The pressure unit of "book" represents the pressure (force per unit area) generated by the weight of a book on the area of the syringe barrel. It is quite convenient to express pressures used in this experiment in units of "books"; however, the pressure will also be converted to the more conventional "torr".

You will obtain volume readings when zero, one, two and three books of the same size and weight are stacked on the syringe. These pressures correspond to relative pressures of zero, one, two and three "books". The pressure of the gas sample must be obtained as total pressure or absolute pressure. This can be obtained graphically by utilizing the relation between pressure (absolute pressure) and volume determined by Robert Boyle. This relation, known as Boyle's Law, can be stated mathematically as:

$$P \, \alpha \, \frac{1}{v} \quad \text{or} \quad P = \frac{k}{V}$$

This shows that a graph of P vs $\frac{1}{V}$ should be linear.

When you have obtained relative pressure and volume readings, calculate 1/V for each reading. For example, a volume reading of 28.1 ml gives a 1/V value of (1/28.1) or .0356. Units on 1/V would be 1/ml or ml^{-1}. Prepare a graph of 1/V on the y-axis vs. relative pressure in "books" on the x-axis. Draw a straight line through the data points and extend the line to 1/V equal to zero. At this point 1/V equals zero, V equals infinity, the relative pressure is a negative number of books and the absolute pressure is zero. Now it is possible to convert each relative pressure reading (zero to three "books") into absolute pressure using "books" as the unit of pressure. Next, read the barometric pressure in torr, and remember that barometric pressure corresponds to a relative pressure of zero books. These ideas can be combined to enable the absolute pressure to be expressed in units of torr. Finally, you will calculate the value of V • P for each experimental point. The volume, V, will be expressed in ml, and the absolute pressure, P will be expressed in torr. Boyle's Law indicates P • V should be constant.

50

This data treatment is complex. Perhaps you will find this example useful in showing you how to treat your data. Assume the data obtained is: zero books, 28.1 ml; one book, 22.6 ml; two books, 19.2 ml; three books, 16.8 ml; barometric pressure, 681 torr. The data and numbers calculated from the data are tabulated below:

Volume, V (ml)	1/V (ml^{-1})	relative pressure (books)	absolute pressure (books)	absolute pressure, P (torr)	V • P (ml • torr)
∞	0	−4.15	0	0	—
28.1	.0356	0.00	4.15	681	1.91×10^4
22.6	.0442	1.00	5.15	845	1.91×10^4
19.2	.0521	2.00	6.15	1009	1.94×10^4
16.8	.0595	3.00	7.15	1173	1.97×10^4

The graph of 1/V vs. relative pressure from the data is shown in Figure 7-III. Extrapolation of the line gives 1/V = O at a relative pressure of −4.15 books. This can be used to calculate an absolute pressure of

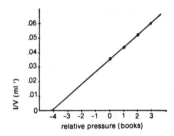

Figure 7-III. Relative pressure vs. I/V

4.15 books or 681 torr for the relative pressure of zero books. From these values, the other numbers can be easily calculated. Notice that V·P is relatively constant.

In order to obtain accurate readings in this experiment, you must carefully clean and lubricate the syringes. Each syringe must have a continuous film of fluid between barrel and plunger. The fluid must have the correct viscosity (thickness) to prevent leaks and to permit easy movement. Follow the recipe for mixing water and glycerol exactly to give the correct type of fluid. Work the barrel and plunger to spread the fluid into a continuous film.

Procedure

Charles's Law

1. Assemble the apparatus shown in Figure 7-I. Use a 50 ml erlenmeyer flask, a 10 ml syringe and a stopper which is completely penetrated by an open hypodermic needle.

2. Clean and dry the barrel and plunger of the syringe. Add two drops of glycerol and two drops of water to the plunger spread the liquid with your finger. Assemble the syringe. Work the plunger back and forth in the barrel, and rotate it slowly, until a continuous, transparent film is produced.

3. Seal the apparatus with 1.5 ml of air in the syringe. Cool the water in the beaker with ice until the plunger is near (but not exactly on) the zero mark. Hold the temperature within 1° for five minutes. Make sure the plunger is not stuck and has reached its "correct" position before each read-

ing. Record the temperature (°C) and the volume of air in the syringe. Heat the water in the beaker about ten degrees. Keep the temperature constant for five minutes to make sure the air sample has reached the same temperature. Check the plunger and record the temperature and volume. Repeat the process until you have obtained readings near 10 ml in the syringe.

4. Note the position of the stopper in the flask. Fill the flask with water to the level of the stopper. Measure the volume of this water by pouring it into a graduated cylinder. This volume corresponds to the volume of air in the flask. The total volume of air is the volume of air in the flask plus the volume of air in the syringe. Prepare a graph of total volume of air on the Y - axis verses temperature. Use the graph to obtain the Celsius temperature where the total volume would be zero, and the value of z where: absolute temperature = Celsius temperature + z. Calculate (total volume/absolute temperature) for each of your readings. Use your absolute temperature scale (your z value) in these calculations.

Boyle's Law

5. Clean and dry a 30 ml syringe. Follow the directions in Step 2 to lubricate it. Assemble the apparatus as shown in Figure 7-II so that 28 or 29 ml of air is trapped. The hypodermic needle should be imbedded in a stopper and should be plugged up. Make sure the clamp holds the syringe vertical; the clamp should be snug but not so tight that it binds the plunger in the barrel.

6. Move the plunger up and down to be sure it is at the "correct" position. Read the volume.

7. Obtain three books of the same weight. Textbooks containing no extra marks, paper or moisture would be ideal. Any three objects of the same weight could be used. Balance one book on the plunger; make sure the plunger is free and in its "correct" position. Read the volume. Repeat with two books and with three books stacked on the plunger. Obtain the barometric pressure reading from your Teaching Assistant. Calculate 1/V for each reading. Prepare a graph of 1/V on Y-axis and relative pressure in books on the X-axis. Draw a straight line through your data points. Extend this line to 1/V equal to zero and read this relative pressure. The relative pressure in books (a negative number) at 1/V equal to zero corresponds to an absolute pressure of zero. Use this fact to complete the data table in a manner similar to the example. The absolute pressure in torr can be obtained from the corresponding absolute pressure in "books", by multiplying each reading by a constant. This constant can be obtained by noting that zero relative pressure corresponds to some absolute pressure (in books) and this is the same as barometric pressure (in torr).

52

Name_____ Section_____

Station_____ Date_____

Gas Laws

Charles's Law   1   2   3   4   5   6   7   8   9

Temperature, °C   ___ ___ ___ ___ ___ ___ ___ ___ ___

Volume of gas in Syringe, ml   ___ ___ ___ ___ ___ ___ ___ ___ ___

Volume of gas, total, ml   ___ ___ ___ ___ ___ ___ ___ ___ ___

Volume of water in flask = _____ ml = volume of air in flask

Celsius temperature where total volume would be zero is_____°C.

Absolute temperature = Celsius temperature + _____ degrees.

Show your calculation of [total volume/absolute temperature] below.

Conclusion:

53

Boyle's Law

Data — zero books, _____ ml; one book, _____ ml; two books, _____ ml;

three books, _____ ml; barometric pressure, _____ torr.

Volume, V (ml)	1/V (ml^{-1})	relative pressure (books)	absolute pressure (books)	absolute pressure (torr)	V · P (ml · torr)
∞	°	_____	zero	zero	_____
_____	_____	zero	_____	_____	_____
_____	_____	one	_____	_____	_____
_____	_____	two	_____	_____	_____
_____	_____	three	_____	_____	_____

Graph of 1/V versus relative pressure:

1/V equal to zero at relative pressure of _____ "books."

At relative pressure of zero books, absolute pressure is _____ "books" or _____

torr. To convert "books" into torr, multiply by_____.

Conclusion:

54

Experiment 8
Measurement of the Universal Gas Constant

Object

In this experiment you will directly measure the value of the universal gas constant, R. The experiment involves the measurement of the pressure, volume, temperature and number of moles of a sample of oxygen gas, O_2.

Background

The ideal gas equation is usually written in the form:

$$PV = nRT$$

In the equation, P represents pressure, V represents volume, n represents number of moles, R is the universal gas constant and T is the absolute (Kelvin) temperature. The equation describes any gas sample with a single value of R. Rearranging the equation into the form:

$$R = \frac{PV}{nT}$$

shows that a value of R can be calculated from the measurements of P, V, n and T for a gas sample.

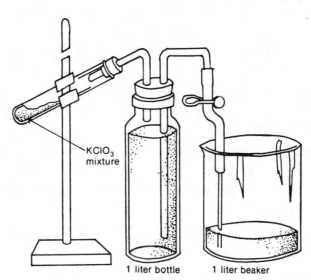

Figure 8-1. Apparatus to measure R

The apparatus shown in Figure 8-I is utilized in the experiment. The test tube contains some potassium chlorate. $KClO_3$, which is decomposed by heating into oxygen gas, O_2, and potassium chloride, KCl. The weight of oxygen produced is determined by weighing the test tube before and after the heating. From the weight of O_2, the moles of O_2 can be obtained. The O_2 sample is collected in the 1ℓ bottle and in the test tube. Notice that this gas is in contact with water. The volume of the O_2

55

produced is determined by measuring the volume of liquid water pushed out of the apparatus. The temperature of the apparatus is measured to obtain T.

The pressure of the O_2 sample is determined in the following way. The water levels in the bottle and beaker are equalized when the pinch clamp is put in place. This insures that the pressure of gases inside the bottle is equal to the pressure outside the apparatus, i.e. barometric pressure. The pressure of the O_2 gas, if it were dry (no water), would be less than barometric pressure. The equation:

$$P_{bar} = P_{O_2} + P_{H_2O}$$

shows the relationship between barometric pressure, P_{bar}; the pressure of the dry O_2 sample, P_{O_2} and the pressure due to water vapor, P_{H_2O}. You may assume that the P_{H_2O} is equal to the vapor pressure of water.

You should compare your measured value of R with the accepted value. Measuring P in torr, V in liters, n in moles and T in degrees Kelvin, the accepted value of R is 62.4 $\dfrac{\ell\text{-torr}}{\text{mole-deg}}$.

Procedure

1. Assemble the apparatus shown in Figure 8-I. Be especially careful when inserting glass tubing into rubber stoppers. Refer to the Introduction for the correct procedures (lubricate the glass and wrap it with a towel). Failure to observe sufficient care at this point will result in broken glass, a cut hand, and your own blood everywhere. Coating the apparatus with blood is definitely NOT recommended!

2. Make sure that the tube connecting the bottle and the beaker is completely filled with water. Attach the pinch clamp. Fill the bottle with water and leave a small amount in the beaker.

3. Remove the test tube from the apparatus. Weigh it. Place about 3.0 to 3.2 g of $KClO_3$ in the test tube. Make sure that none of this material comes in contact with the stopper. Weigh the test tube accurately. The $KClO_3$ contains some KCl and some MnO_2 to insure smooth decomposition.

4. Replace the test tube in the apparatus. Remove the pinch clamp. Make sure the apparatus is not leaking. Equalize the water levels in the beaker and the bottle and replace the pinch clamp. Empty the beaker. *Remove the pinch clamp.*

5. Gently heat the $KClO_3$ mixture with a bunsen burner flame. Do not permit bubbles to enter the bottle from the beaker. Keep the tip of the tube under water. Do not overheat the $KClO_3$— if white smoke appears over the $KClO_3$ discontinue heating. When 600-800 ml of water have been collected in the beaker, discontinue heating.

6. When the test tube has cooled to room temperature, equalize the water levels and attach the pinch clamp. Measure the temperature of the apparatus. Measure the volume of water collected in a one liter graduated cylinder. Record the barometric pressure.

7. Remove the test tube from the apparatus and weigh it accurately on the same balance used in Step 3.

8. From your data, calculate the number of moles of O_2. Calculate the volume and temperature of the O_2 sample. For the calculation of the pressure of the dry O_2, you will need to look up the vapor pressure of water. Calculate the value of R. Calculate the per cent error in your R value.

56

Name_____ Section_____

Station_____ Date_____

Measurement of the Universal Gas Constant

Weight of empty test tube,_____g

Weight of tube plus $KClO_3$ mixture, _____g

Weight of tube plus residue, after heating,_____g

Weight of O_2,_____g; moles of O_2,_____moles.

Show your calculations below.

Volume of water = Volume of O_2 = _____ml or _____ℓ.

Temperature of apparatus,_____°C or_____°K.

Vapor pressure of water, _____torr.

Barometric pressure, _____torr.

Pressure of dry O_2 sample,_____torr. Show your calculations below.

Value of R = _____ $\dfrac{\text{torr-}\ell}{\text{mole-deg}}$ Show your calculations below.

Per cent error in experimental R value is _____%. Show your work below.

57

Experiment 9
The Molecular Weight of a Volatile Liquid

Object

The molecular weight of an unknown volatile liquid will be determined by a measurement of the density of the vapors of the liquid.

Background

John Baptiste Andre Dumas (1800-1884) made several contributions to chemistry. Today's experiment is based on a method he developed for determination of molecular weights by measurement of the vapor density. The equations which form a basis for the calculations are the ideal gas equation

$$PV = nRT$$

where P is pressure, V is volume, n is number of moles, R is the universal gas constant and T is temperature on the Kelvin scale. Since the number of moles, n, is also given by $n = w/MW$ where w is the weight of a sample and MW is the molecular weight, we can also write

$$PV = \frac{w}{MW}RT, \text{ or, rearranging, } MW = \frac{w\,RT}{V\,P}.$$

The last equation shows that the molecular weight of a gas can be calculated from measurements of the weight, temperature, volume, and pressure of a sample of the gas.

The weight of a gas sample is very difficult to measure. The volume, temperature and pressure of the vapors of your unknown will be measured at the boiling point of water (about 96°C or 369°K at Lubbock). Under these conditions, the flask will be completely filled by vapors of the unknown but no liquid will be present. Unknowns are chosen from liquids having boiling points at least 20°C below the boiling point of water. To avoid the problem of weighing a gas however, the vapors will be condensed back to a liquid before the weight is measured. The measurement of the weight will be done with the flask at room temperature. The unknown will be present as a liquid under these conditions.

The universal gas constant has the same value for all substances. It has a value of 1.99 cal/mole deg or 0.0821 ℓ atm/mole deg or 82.1 ml atm/mole deg. When R is used in the equation at 0.0821 ℓ atm/mole deg it is easiest to use w in grams, T in Kelvin degrees, V in ℓ and P in atm. Units will cancel so that MW will be calculated in units of g/mole. In other words MW is the number of grams in one mole of the substance or the molecular weight of the substance.

Experimental Procedure

1. Obtain a piece of aluminum foil. Place it between two sheets of paper and punch a pinhole in the center of the foil.

2. Obtain a clean, dry 100-ml volumetric flask. Select a balance; check its zero point; and weigh the flask and foil to ±0.001 g.

59

3. Pour about 6 ml [*3 squirts*] of your unknown liquid into the flask. Cover the neck of the flask with the foil. Make sure that the cap is secure on the neck, but that vapors can escape through the pinhole.

4. Place the flask in boiling water. Use a water bath in the hood; a beaker of boiling water in the hood, or a beaker of boiling water at your desk.

5. After all the liquid in the flask has evaporated, [*take it out*] ~~continue boiling for 5 minutes.~~ Record the temperature to ±0.1°. Record the barometric pressure.

6. [*Put under water faucet*] Remove the flask from boiling water and immerse it in cold water. Dry the outside of the flask carefully. Make sure that no water has condensed under the foil, but do not remove the foil from the neck of the flask.

7. Weigh the flask, foil and condensed vapors using the same balance. Be sure to check the zero point of the balance.

8. [*Repeat.*] Dry the flask. Obtain another piece of foil (if necessary) and perform the experiment again.

9. Dry the flask again. Fill the flask with distilled water to the 100 ml mark. Measure the volume of water necessary to completely fill the flask. Your instructor will have burets set up in the hood to make this measurement. [*Use graduated cylinder*]

[*Weigh flask – boiling stone – tin foil and/or rubberband*]

[*put 3 squirts*]

Experiment 10
Molecular Weight of a Solute

Object

In todays' experiment you will measure the depression in the freezing of diphenyl caused by dissolving naphthalene in the biphenyl. You will then measure the depression in freezing point caused by an unknown substance and calculate the molecular weight of the unknown substance.

Background

The freezing point is defined as the temperature where solid begins to form in a liquid. If a substance is pure (pure H_2O for example) it will have a constant temperature until all the liquid has been converted into solid (0°C or 32°F, where water is converted into ice). If a substance is a mixture, however, it will have a lower freezing point. Pure water freezes at 0.00°C, but a mixture containing 1 mole glucose and 1 kg water will freeze at-1.86°C. A mixture containing 1 mole glucose and 500 g water will freeze at –3.72°C. The depression of the freezing point depends on the concentration. It is most convenient to express the concentration in molality, m or moles of solute per kg of solvent; because this concentration scale is independent of temperature. The freezing point depression, ΔT, can be calculated from the equation:

$$\Delta T = Km$$

In this equation, m represents the molality of the solution, and K is the molal freezing point constant. It is a property of the solvent. For water the value of K is 1.86 deg/m, but for diphenyl the value of K is 8.0 deg/m.

Measurement of the freezing point depression can be used to obtain molecular weight. If you dissolve 1.38g of ethanol in 60.0g water, the solution will have a freezing point of –0.93°C. We can calculate that the solution has a concentration of 0.50 molal, or the solution contains 0.5 moles of ethanol per kg of water. Since 1.38g ethanol per 60.0g water is equivalent to 23g of ethanol per kg of water, we can see that 23g of ethanol is 0.50 moles of ethanol. One mole of ethanol is 46g or the molecular weight of ethanol is 46.

Water is a very common solvent in chemistry. For measurements of molecular weight by freezing point depression, we can list the following disadvantages: (1) freezing points are inconvenient since they are below 0°C (2) only polar substances will dissolve in water (3) K is quite low, only 1.86 deg/m. Diphenyl, C_6H_5-C_6H_5 or $C_{12}H_{10}$ is a very convenient solvent for measurements of molecular weights. It has a convenient freezing point of about 60°C; it dissolves non-polar substances readily. For diphenyl, K is 6.0 deg/m; and is therefore higher than for water.

Apparatus for measurement of the freezing point is shown in Figure 10-I. Diphenyl or a diphenyl solution is placed in the test tube. Water in the beaker is heated until the solution is a liquid. The water and diphenyl is stirred as cooling occurs. Time and temperature readings are recorded at 30 second intervals. Typical plots of the experimental points are shown by the solid lines in Figure 10-II. The upper curve on the graph represents pure solvent. The pure solvent cools rapidly at first; then as solid begins to form the temperature remains constant. Supercooling may occur when the liquid is below the freezing point (before solid appears). As solid appears, the mixture will warm up to the freezing point. The temperature

63

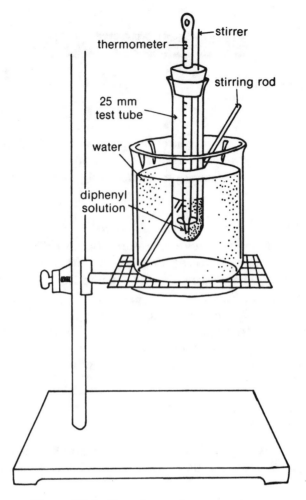

Figure 10-I. Freezing point apparatus

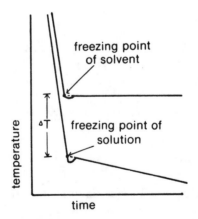

Figure 10-II.

64

should remain constant during solidification of a pure substance. The freezing point is the intersection of the two straight lines, and is indicated on the diagram. The lower curve represents the solution. Notice that supercooling is indicated in this also. The temperature decreases slowly during solidifcation, since the concentration of solute in the liquid is increasing as solvent turns into solid. The freezing point is indicated on the drawing. It is obtained by extending the straight line part of the solidification curve back to the liquid curve. The freezing point depression is the difference in the freezing points, and is indicated as ΔT on the graph.

In today's experiment, you will measure the freezing point of diphenyl. You will dissolve naphthalene, $C_{10}H_8$ into the diphenyl and measure the freezing point of the solution. You can calculate the molecular weight of naphthalene from your data by combining (a) the grams of naphthalene per kg of diphenyl from the weights mixed together, and (b) the molality of the solution from the ΔT you measure and the K value of 6.0 deg/m for diphenyl. You will then calculate the per cent error in your experimental molecular weight value. You will then prepare a solution containing an unknown substance dissolved in diphenyl. You will measure the freezing point of the solution. From this data you will calculate the molecular weight of the unknown substance.

Procedure

1. Assemble the apparatus shown in Figure 10-I. Be careful when inserting the thermometer in the rubber stopper. Lubricate the thermometer, cover it with a towel when pushing it through, and don't push too hard!

2. Accurately weigh the test tube. Add about 15g diphenyl and accurately reweigh. Reassemble the apparatus and heat the water to about 75°C. When all the diphenyl has melted, allow the temperature to drop slowly. Record temperature readings (to ±0.1 degree) at 30 sec. intervals and note the appearance of solid diphenyl in the test tube. Obtain readings for four minutes after the solid appears. Stir both the water and the diphenyl during these readings. Graph your results and determine the freezing point of the pure diphenyl. Write out your comments on the temperature during solidification.

3. Weight a small piece of paper (10cm x 10 cm). Pour about 3g of naphthalene on the paper and reweigh. Carry the paper carefully and then pour *all* the naphthalene into the test tube containing the diphenyl. Heat the apparatus so that the contents of the test tube are melted. Mix the solution well. Permit the cooling to occur. When the temperature reaches 72°C, begin recording temperature (to ±0.1 deg) at 30 sec. intervals. Stir the diphenyl solution and the water well during these readings. Note the point where solid forms, and obtain readings for an additional four minutes.

4. Graph your results and determine the freezing point of the naphthalene solution. Write out your comments on the temperature during solidification. Calculate ΔT (freezing point depression) and then calculate the molality of the solution. From the weights of naphthalene and diphenyl, calculate the grams of naphthalene per kg of diphenyl. Calculate the weight of one mole of naphthalene, from your experimental data. Calculate the % error in your value, assuming the correct formula for naphthalene is $C_{10}H_8$.

5. Heat the apparatus until the water is nearly boiling. Remove the thermometer and stirrer, and while they are still hot, clean them off with paper towels. Pour the contents of the test tube into the special solid waste can in the laboratory. Reheat the tube in boiling water if necessary.

6. When the test tube, thermometer, and stirrer are clean and dry, prepare a solution containing the unknown. Weigh the test tube, add about 15g diphenyl and reweigh. Add about 3g of un-

known and reweigh. Assemble the apparatus and heat the unknown solution to melt it completely. Stir well, and when the temperature drops to 72°, begin recording temperature (to ±0.1 deg) at 30 sec intervals. Note the appearance of solid and continue making readings for four minutes after this time.

7. Heat the water bath to boiling. Clean the test tube, thermometer and stirrer as before.

8. Graph your results and determine the freezing point of the unknown solution. Calculate ΔT (the freezing point depression) and the molality of the solution. Calculate the grams of unknown per kg of diphenyl. Calculate the molecular weight of the unknown.

Name_____   Section_____

Station_____   Date_____

Molecular Weight of a Solute

Weight of empty test tube, _____ g; weight of diphenyl + test tube,_____ g.

Weight of diphenyl, _____ g;

Temperature readings at 30 sec. intervals for pure diphenyl:

_____ _____ _____ _____ _____ _____

_____ _____ _____ _____ _____ _____

Solid first appeared at temperature of _____°C.

Comments on temperature during solidification.

Weight of paper,_____ g; weight of paper plus naphthalene,_____ g.

Weight of naphthalene,_____ g; _____ g of naphthalene per kg of diphenyl.

Show your work below.

Temperature reading for naphthalene solution at 30 sec. intervals:

_____ _____ _____ _____ _____ _____

_____ _____ _____ _____ _____ _____

Solid first appeared at temperature of_____°C.

Comments on temperature during solidification.

Freezing point of pure diphenyl, from graph, is_____°C.

Freezing point of naphthalene solution, from graph, is_____°C.

Freezing point depression of naphthalene is_____°C.

Molality of naphthalene, from freezing point depression, is _____moles/kg.

Show your work below.

Grams of naphthalene per mole of naphthalene, from your data, is_____.

Correct molecular weight for $C_{10}H_8$ is_____.

% error in experimental value is_____. Show your work below.

Weight of empty test tube,_____ g; weight of test tube plus diphenyl,_____g.

Weight of test tube, diphenyl and unknown,_____ g.

Solution contains_____g unknown and_____g diphenyl;

Solution contains_____g unknown per kg diphenyl.

Temperature readings for unknown solution at 30 sec. intervals:

_____ _____ _____ _____ _____ _____

_____ _____ _____ _____ _____ _____

Solid first appeared at temperature of_____°C

Freezing point of unknown solution, from graph, is_____°C.

Freezing point depression of unknown solution is_____°C.

Molality of unknown solution is_____moles/kg diphenyl.

Molecular weight of unknown is_____g/mole. Show your work below.

Name _____   Section _____

Station _____   Date _____

Molecular Weight of a Solute

Experiment 11
Coefficients in a Chemical Equation

Object

In this experiment you will measure the moles of hydrogen produced, per mole of magnesium, from the reaction of magnesium with an acid. The moles of magnesium will be determined from the weight of magnesium. The moles of hydrogen will be determined from the pressure, volume and temperature of the hydrogen produced.

Background

Magnesium metal readily reacts with acids to produce hydrogen and a magnesium salt. The equation may be written in the form:

Magnesium Metal + acid $\longrightarrow$ magnesium salt + hydrogen gas

You will measure the moles of magnesium reacted in the experiment and the moles of hydrogen produced in the experiment. The ratio of these numbers is the same as the ratio of the coefficients in the balanced chemical equation.

The apparatus used in this experiement is diagramed in Figure 11-I. A weighed magnesium sample is suspended by a thread in the erlenmeyer flask. The flask is tilted so that the magnesium is reacted

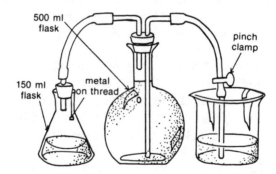

Figure 11-I. Apparatus to measure hydrogen generated by magnesium

with the acid solution in the erlenmeyer flask. As hydrogen gas is produced by the reaction, water is pushed out of the apparatus and caught in the beaker. When the reaction is finished, the pressure, volume and temperature of the hydrogen gas is measured. The number of moles of hydrogen can be calculated from the ideal gas equation:

$$PV = nRT \text{ or } n = \frac{PV}{RT}$$

71

Recall that P represents pressure, V represents volume, R is the universal gas constant, and T is the Kelvin temperature. The number of moles of gas is n. If you utilize $R = 62.4 \frac{torr-\ell}{mole-deg}$ then the pressure must be expressed in torr, the volume in liters and the temperature in degrees Kelvin. The measurement of T is straightforward. The volume of water collected is equal to the volume of hydrogen produced. The gas sample is collected, over water, at barometric pressure. The gas pressure must be corrected for the pressure due to water. Refer to Experiment 8 for further details on this point.

A good deal of confusion occurred in Chemistry at one time regarding the molecular formula for hydrogen. It is now well established that hydrogen gas contains H_2 molecules. Utilizing hydrochloric acid, HCl, to react with magnesium, we can write the equation:

$$aMg + bHCl \longrightarrow cMagnesium\ chloride + dH_2(gas)$$

In the experiment, the ratio of d/a will be measured. Once this ratio is determined, you should be able to complete the equation. You must choose a formula for magnesium chloride, and assign coefficients a, b, c, and d to make the final equation obey the Law of Conservation of Mass. If time permits, you will repeat the experiment using sulfuric acid, H_2SO_4, instead of hydrochloric acid. The equation for this reaction can be written:

$$eMg + fH_2SO_4 \longrightarrow gMagnesium\ sulfate + hH_2(gas)$$

Procedure

1. Assemble the apparatus shown in Figure 11-I. Be especially careful when inserting the glass tubing into the rubber stoppers. Refer to the Introduction for instruction on the correct procedure.

2. Make sure the tube connecting the florence flask and beaker is completely filled with water. Attach the pinch clamp. Pour 10 ml of concentrated HCl and one drop of 0.01 M $Cu(NO_3)_2$ into the erlenmeyer flask. Cut a piece of Mg ribbon into a length which will weigh 0.2 to 0.3 g. Accurately weigh the Mg sample. Fold the Mg into a small wad and tie a short length of thread to the metal. Hang the Mg inside the erlenmeyer flask. (Alternatively, small pieces of Mg metal can be tied in a cheesecloth bag which can be hung with thread inside the flask).

3. Make sure the apparatus does not leak. Remove the pinch clamp and equalize the water levels in the beaker and the florence flask. Attach the clamp and then empty the beaker. Replace the empty beaker and remove the pinch clamp. Tilt the erlenmeyer flask so that the Mg comes in contact with the HC1 solution. If the reaction becomes too rapid, place the flask in an upright position for a few seconds. When all the metal has dissolved, permit the solutions to cool a few minutes.

4. Record the temperature. Equalize the liquid levels in the beaker and florence flask and then attach the pinch clamp. Measure the volume of water collected in a graduated cylinder. Record the barometric pressure. Look up the vapor pressure of water and calculate the pressure of dry H_2 gas.

5. Calculate the moles of magnesium, moles of hydrogen and moles of hydrogen per mole of Magnesium for your experiment. Show that only one of the following formulae for magnesium chloride can be used to balance the equation in a way which is consistent with your data: $MgCl$, Mg_2Cl, $MgCl_2$, Mg_2Cl_3. Write out the balanced equation.

6. If time permits repeat the experiment with 10 ml of 6 M H_2SO_4 and one drop of 0.01 M $CuSO_4$ in the erlenmeyer flask.

7. Treat your data in the same way. Determine which formula for Magnesium sulfate is consistent with your data: $MgSO_4$, Mg_2SO_4, $Mg(SO_4)_2$ or $Mg_2(SO_4)_3$. Write out the equation.

Name_____   Section_____ _____

Station_____   Date_____

Coefficients in a Chemical Equation

For the reaction: Mg + HCl

Weight of Mg,_____ g; moles of Mg,_____ moles. Show your work.

Temperature of gas sample,_____°C or_____°K.

Volume of gas sample,_____ ml or_____ ℓ.

Barometric pressure,_____ torr.

Vapor pressure of water at_____°C is_____torr.

Pressure of dry sample of H_2 is_____torr. Show your work below.

Moles of H_2 produced in the reaction is_____ moles. Show your work below.

Moles of H_2/moles of Mg =_____= ratio of coefficients in the equation.

The equation for Mg + HCl can be written as:

1.0 Mg +_____HCl $\rightarrow$_____Magnesium chloride +_____H_2 (gas)

The formula for magnesium chloride must be_____. Show your reasons below.

Show other formulae are not consistent.

73

For the reaction: $Mg + H_2SO_4$

Weight of Mg,_____ g; moles of Mg,_____ moles. Show your work.

Temperature of gas sample,_____°C or _____°K.

Volume of gas sample,_____ ml or_____ ℓ.

Barometric pressure, _____ torr.

Vapor pressure of water at_____°C is_____ torr.

Pressure of dry sample of H_2 is_____torr. Show your work below.

Moles of H_2 produced in the reaction is_____moles. Show your work below.

Moles of H_2/moles of Mg =_____ = ratio of coefficients in the equation.

The equation for $Mg + H_2SO_4$ can be written as:

1.0 Mg +_____$H_2SO_4 \longrightarrow$_____Magnesium sulfate +_____H_2 (gas)

The formula for Magnesium sulfate must be_____. Show your reasons below. Show other formulae are not consistent.

Write the complete equation below.

74

Experiment 12
Activity Series

Object

The relative reactivities of several metals in chemical reactions will be studied.

Background

The Periodic Table can be divided into two main parts: the *metals* are on the left side and the *nonmetals* on the right side. A jagged line (generally along the area from B to Si to As to Sb to Po) separates these two "halves" of the Periodic Table. The elements along the jagged line have some properties of both the metals and nonmetals; these elements are referred to as *metalloids*. The metals have physical properties with which most of us are familiar. they are lusterous, they conduct heat and electricty, they are malleable (can be pounded flat without shattering) and ductile (can be drawn out into a fine wire without breaking), and they are solids at room temperature (except for mercury). The metals also have equally distinctive chemical properties.

Metals tend to lose electrons easily, i.e. they have low ionization potentials. This reaction can be represented as:

$$\text{Metal} \longrightarrow \text{Metal}^{n+} + n\ e^-$$

In other words, metals tend to be easily oxidized (which also means they are good reducing agents). Four general reactions where a metal can transfer electrons are: a) reaction with water, b) reaction with oxygen, c) reaction with acids, and d) reaction with other metal ions. These four reaction types can be illustrated by using a very reactive metal, sodium, which is in Group I (the alkali metals) of the Periodic Chart.

Reaction with Water. Sodium reacts with water with considerable violence (a very exothermic reaction) to produce sodium hydroxide and hydrogen gas.

$$2Na_{(s)} + H_2O_{(1)} \longrightarrow 2\ NaOH_{(aq)} + H_{2(g)}$$

The reaction can be divided into two half reactions:

$$2Na_{(s)} \longrightarrow 2\ Na^+_{(aq)} + 2e^-, \text{ oxidation half reaction}$$

$$2e^- + 2H_2O_{(1)} \longrightarrow 2\ OH^-_{(aq)} + H_{2(g)}, \text{ reduction half reaction}$$

The more reactive metals will react with water at room temperature, the less reactive ones will only react with water at elevated temperatures.

Reaction with Oxygen. Sodium spontaneously reacts with oxygen to produce the oxide.

$$4Na_{(s)} + O_{2(g)} \longrightarrow 2\ Na_2O_{(s)}$$

75

Sodium is being oxidized from the metal to the metal cation; oxygen is behaving as an oxidizing agent and is being reduced to the oxide anion O^{-2}.

Reaction with Acids. Most metals will react with acids such as hydrochloric acid.

$$2\ Na_{(s)} + 2\ HCl_{(aq)} \longrightarrow 2\ NaCl_{(aq)} + H_{2(g)}, \text{ or more simply,}$$

$$2\ Na_{(s)} + 2\ H^+_{(aq)} + 2\ Cl^-_{(aq)} \longrightarrow 2\ Na^+_{(aq)} + 2\ Cl^-_{(aq)} + H_{2(g)}$$

Since $2\ Cl^-_{(aq)}$ occurs on both sides of the equation, chloride is not entering into the chemical reaction; chloride is merely a *spectator* ion which may be omitted from the net ionic equation.

$$2\ Na_{(s)} + 2\ H^+_{(aq)} \longrightarrow 2\ Na^+_{(aq)} + H_{2(g)}$$

or, more simply,

$$2\ Na + 2\ H^+ \longrightarrow 2\ Na^+ + H_2$$

Reaction with Other Metal Ions. Since sodium is such a reactive metal that it reacts with water, we will choose another metal to illustrate the reaction of a metal with the ion of another metal.

$$Cu_{(s)} + 2\ AgNO_{3(aq)} \longrightarrow Cu(NO_3)_{2(aq)} + 2\ Ag_{(s)}$$

or

$$Cu_{(s)} + 2\ Ag^+_{(aq)} + 2\ NO^-_{3(aq)} \longrightarrow Cu^{2+}_{(aq)} + 2\ NO^-_{3(aq)} + 2\ Ag_{(s)}$$

Again, nitrate NO_3^- is a spectator ion, so the equation net ionic reaction can be written as:

$$Cu + 2\ Ag^+ \longrightarrow Cu^{2+} + 2\ Ag$$

This type of reaction is also called a "single replacement reaction" in some textbooks.

The activity series is a listing of metals according to their reactivity as reducing agents. That is, the more "active" a metal is in its elemental (uncombined) state, the more likely it is to donate electrons to a positive ion (cation) in solution, and thus replace it in solution. In the equation above, copper is more "active" than silver, and thus copper replaces silver ions in solution. Hydrogen is usually included in the activity series. Metals more active than hydrogen are called "active metals", and metals less active than hydrogen are called "inactive metals". One expects single replacement reactions to occur between a metal in its elemental state with a positive metal ion lower on the activity series: the more active metal replaces the less active metal ion in solution.

Procedure

Work in pairs.

1. In the first part of the experiment, the object is to study the reaction of metals with acid. To each of five test tubes containing about 5 ml of 6 M HCl, add a small piece of a metal. One test tube will have a piece of Mg, one Fe, one Cu, one Zn, and one Al. Observe the test tube carefully and note any reaction such as a gas being produced, any color changes, and whether or not the reaction is slow or vigorous. Write both complete and net ionic equations.

Mg^{2+} , Ni^{2+} , Cu^{2+} , Zn^{2+} , Al^{3+}

no rxn

76

2. In this part of the experiment, the object is to determine whether one metal will replace the other metal ion in solution. To each of five test tubes containing, respectively, several mL of $Mg(NO_3)_2$, $Fe(NO_3)_3$, $Cu(NO_3)_2$, $Zn(NO_3)_2$, and $Al(NO_3)_3$, add a small piece of magnesium metal. Note whether any reaction occurs by observing any color change on the surface of the metal or whether any gas is evolved. Record your observations, and write the net ionic equations, assuming nitrate is only a spectator ion. Repeat the testing with each of the other metals: Fe, Cu, Zn, and Al. Be sure to record all evidence of reaction.

3. Using the information which you are able to put together from your observations in part 2 of this experiment, you should be able to rank the five metals as to their activity. The most active metal will replace all of the other ions, the next most active metal should replace all of the metal ions from solution except the most reactive, etc. Deduce the order of activity, and record your prediction. Answer the other related questions.

$$2\,Al_{(s)} + 6\,HCl_{(aq)} \rightarrow 2\,AlCl_{3\,(aq)} + 3\,H_{2\,(g)}$$

$$net:\ 2\,Al_{(s)} + 6\,H^+_{(aq)} \longrightarrow 2\,Al^{3+}_{(aq)} + 3\,H_{2\,(g)}$$

$$Mg_{(s)} + 2\,HCl_{(aq)} \longrightarrow MgCl_{2\,(aq)} + H_{2\,(g)}$$

$$net:\ Mg$$

Experiment 13
Basics of Acid-Base Titration

Object

This experiment demonstrates how to use a precise analytical technique called titration. The amount of acid in a solution will be determined by titration.

Background

The purpose of a chemical *titration* is to determine the concentration of a substance in solution (or simply the quantity of that substance per unit volume of solution). The most common unit of concentration is *molarity,* M, which is defined as the number of gram molecular weights (moles) per liter of solution. For example the gram molecular weight of sodium hydroxide is 40 g. If 40 g of NaOH is dissolved in sufficient water to make one liter of solution, the concentration of NaOH is exactly 1.00 M (1.00 gram molecular weight or 1.00 mole per liter). If 20 g of NaOH is dissolved to make one liter of solution, the concentration is 0.50 M (one-half of a gram molecular weight or mole per liter). Despite the fact that solutions 2 and 3 in the table were made differently, the concentration of NaOH is the same in each, i.e., 0.50 M. A 1.0 ml sample from each contains exactly the same quantity of dissolved NaOH.

Solution #	g of NaOH	# of moles of NaOH (or # of g.mol.wts.)	Total Volume	Molarity
1	40	40g/(40g per mole)=1.00	1.00 ℓ	1.00 moles/1.00 ℓ=1.00 M
2	20	20g/(40g per mole)=0.50	1.00 ℓ	0.50 moles/1.00 ℓ=0.50 M
3	40	40g/(40g per mole)=1.00	2.00 ℓ	1.00 moles/2.00 ℓ=0.50 M
4	40	40g/(40g per mole)=1.00	0.50 ℓ	1.00 moles/0.50 ℓ=2.00 M
5	80	20g/(40g per mole)=2.00	1.00 ℓ	2.00 moles/1.00 ℓ=2.00 M

In this experiment the concentration of an unknown acid in a solution will be determined by measuring the amount of acid solution required to completely react with a known quantity of base. The acid is hydrochloric acid, HCl, which is an important component in stomach digestive juices. The base is sodium hydroxide, commonly known as lye. The reaction can be written as:

$$HCl + NaOH \longrightarrow H_2O + NaCl$$

or more simply

$$H^+ + OH^- \longrightarrow H_2O$$

One H^+ ion reacts with exactly one OH^- ion. Equal volumes of acid and base of equal concentration will exactly react or *neutralize* each other. Thus 1.0 ml of 1.0 M base neutralizes exactly 1.0 ml of 1.0 M acid.

81

If the base were 2.0 M, then 1.0 ml would require 2.0 ml of 1.0 M acid to neutralize it, since the base is twice as concentrated as the acid. The general relationship for neutralization is

$$(\text{volume of acid}) \times (\text{Molarity of acid}) = (\text{volume of base}) \times (\text{Molarity of base})$$

In the experiment you will take a known quantity of acid of unknown concentration. You will then add exactly the volume of base required to neutralize the acid. The concentration (molarity) of the base is known (it will be given to you), and, using the above equation, you can then calculate the concentration (molarity) of the unknown acid. The "end point of the titration" when the acid and base exactly neutralize each other is determined by use of the indicator phenolphthalein. The indicator is colorless in acid solution and is red in base solution. The end point of the titration when the base just neutralizes the acid is when the solution turns a barely detectable pink color.

Volumes of solutions can be measured easily to an accuracy of ±0.02 ml by using a buret. To read a buret, have your eye level with the *bottom* of the curved surface of the liquid (meniscus). To check your eye level, use the rings marked on the buret. If the ring nearest the meniscus appears as a straight line, then your eye is at the proper height to read the buret. The volume should be read to *two decimal places.* The second decimal place (hundreths) is estimated by mentally dividing the space between the two lines into 10 equal parts, and counting the number of these parts required to go from the line just above the meniscus to the bottom of the meniscus.

Experimental Procedure

1. The NaOH solution has been standardized, that is, its concentration is known. Write the concentration on your data sheet. Obtain about 150 ml of the NaOH in a clean dry beaker. Obtain about 150 ml [100 ml] of the unknown HCl in a clean dry beaker.

2. Set up two clean burets. Clean the burets until water does not cling to the inside of the buret as it is drained. It is not necessary to dry the buret after cleaning. Just rinse with several portions of distilled water.

3. Rinse one buret with two 5-ml portions of the standard NaOH solution. Do this by pouring in about 5 ml of NaOH either directly from the beaker or through a funnel. Place your thumb over the top of the buret and invert the buret several times in order to rinse the inside surface. Drain the NaOH into a beaker and discard it. Repeat the procedure with a second 5-ml quantity of NaOH.

4. Repeat step 3 with the other buret, rinsing with two 5-ml quantities of your unknown HCl solution.

5. Fill the *NaOH-cleaned* buret with *standardized NaOH* to slightly above the zero mark and clamp the buret up vertically.

6. Repeat step 5 with your unknown acid, using the HCl-cleaned buret.

7. To remove the air bubbles from the tip of the buret, place your finger over the tip and, with the other hand, squeeze the pinch clamp and move it down over the glass tip. Release the clamp and squeeze the rubber tubing several times to force the bubbles out the top. Move the pinch clamp back in place and rinse your hands.

8. Drain both burets to slightly below the zero mark and read the level, recording the reading to ±0.02 ml. Discard the solution that is drained.

82

9. Drain approximately 25 ml *[handwritten: 10-15 ml]* of the acid into the clean, but not necessarily dry, conical flask. Do not read the acid level at this time. Wait until after the titration is complete. *Add 4 drops of phenolphthalein indicator to the flask.* *[handwritten: or squirt]*

10. Carefully place the flask containing the acid and indicator under the NaOH-containing buret. Slowly add NaOH from the buret to the flask with swirling until the solution in the flask is barely a detectable permanent pink. There should be a one-drop difference between when the solution is colorless and when it is pink (for several minutes). If too much base is added (that is, if you "overshot" the endpoint), add a small amount more acid until the color just disappears. Then slowly add the base until the pink color permanently appears. Again, the color should be barely visible. A white piece of paper under the flask will aid in the color detection.

11. When the proper endpoint is reached, record the final readings on both burets to the nearest ±0.02 ml.

12. Repeat the titration a second time, filling the burets again. Record the initial readings and final readings.

13. Calculate the concentration of your acid for each titration.

14. If the calculated acid concentration from the first and second titrations vary by more than 0.005 M, perform a third titration.

15. Calculate the average acid concentration.

[handwritten: phenol- pthaleia]

[handwritten:]
$$(.1 \ M \ NaOH)(.0108 \ L) = (.01240 \ L)(x \ M \ HCl)$$

$$\frac{.00108}{.01240} = .01240 \ x$$

$$x = .0871$$

Experiment 14
Applications of Acid-Base Titrations

Object

In this experiment you will learn how to perform an acid-base titration. In the first titrations, you will measure the concentration of the base solution. In the second titrations, you will determine the gram equivalent weight of an unknown acid.

Background

The titration experiment is a routine but important procedure in most laboratories. In the acid-base titrations you perform today, an acid sample (in the flask) is titrated with a basic solution from the buret until the end point of the titration is reached. Reaching the end point, without overshooting the end point, is a key feature of any titration. At the end point of a titration, the acid and base are chemically equivalent. This will mean that equivalents of base or moles of hydroxide ion added to the flask are exactly equal to equivalents of acid or moles of hydrogen ion added to the flask.

The gram equivalent weight (GEW) of a base is the amount required to produce one gram-mole of hydroxide ions. Forty grams of sodium hydroxide, NaOH, contains one mole of NaOH. From the formula, you can see that one mole of NaOH will supply one mole of hydroxide ions, OH^-. For an acid, the GEW is defined as the amount which produces one mole of hydrogen ions, H^+. For HCl, the GEW is 36.5 grams. For acetic acid, $HC_2H_3O_2$, the GEW is the same as the molecular weight or 60 g. Even though the acetic acid molecule contains four hydrogen atoms, only one is able to produce H^+ when reacting with a base. The formula of an organic acid is usually written to show the acidic hydrogens first in the formula. Tartaric acid, $H_2C_4O_6H_4$, has a GEW of 75. The molecular weight is 150, but the formula shows that each molecule contains two acidic hydrogen atoms. A mole of tartaric acid, 150 g, can provide two moles of H^+.

For titrations, solution concentrations are most conveniently expressed as normality, N. Normality of a solution is defined as number of GEW per liter solution. When a solution contains 4 g of NaOH per liter of solution, the solution is 0.1 N or 0.1 M. When a solution contains 22.5 g of tartaric acid per liter of solution, the solution is 0.30 N or 0.15 M. When the titration is performed correctly and the end point is obtained accurately, the following equalities may be stated for the chemical system:

$$N_A V_A = N_B V_B$$

In this equation, N_A and N_B are the normalities of the acid solution and the base solution. V_A and V_B are the volumes of the solutions used in the titration. Notice that $N_B \times V_B$ = number of equivalents of base, provided the volume is in liters. In case a sample of a solid acid of weight, W, and a known GEW is titrated to the end point, we can write the equation:

$$V_B N_B = \frac{W}{GEW}$$

In most titrations, an indicator is added to the solution to enable the end point to be detected. Phenolphthalein is used as an indicator in all of the titrations to be performed today. This dye is colorless in acidic medium, but is light pink at the end point. If you overshoot the end point, the dye will be deep pink or red. The secret to accurate titrations is to add the last portion of base in very small amounts. You want to add one drop (or less) of base to cause the pink color to appear in the solution.

You will first prepare a NaOH solution and pour it into the buret. After you make sure that this solution is uniform, you will measure the normality of the base. This process is called, "standardizing the base." The standardization is done by titrating weighed samples of potassium acid phthalate. This acid is a good primary standard because it is stable, easy to keep dry, and has a high GEW value of 204.2. From the weight of potassium acid phthalate, the GEW value, and the volume of NaOH solution to reach the end point, the normality of the basic solution can be calculated. The process of standardizing the base is repeated until suitable value for the normality is obtained.

After the base has been standardized, it is used to measure the GEW of an unknown acid. In this experiment, however, the solid acid is weighed and transfered to a volumetric flask. The solution is diluted to a volume of exactly 100 ml, and 20 or 25 ml portions of the solution are titrated. From the titration data you can calculate the normality of the acidic solution. From this normality, and the weight of acid you dissolved, you can calculate the GEW of the unknown acid.

Procedure

1. Prepare a .012 N NaOH solution. Pour about 48 ml of 1 N NaOH into a 50 ml beaker. Pour this into a 500 ml florence flask and add about 350 ml of distilled water. Mix the solution well, and hereafter, keep the solution from being contaminated. Cover the flask with a dry beaker. Use this beaker to transfer basic solution from the flask to the buret.

2. Clean a buret with soap and water and a buret brush. Rinse with tap water. Drain as well as possible. The buret is clean enough when no water droplets cling to the inner surface. Rinse with three 4 ml portions of distilled water. Be sure to rinse the lower part of the buret (below the glass bead). Rinse the buret with two 3 ml portions of NaOH solution, and discard this liquid. Set up the buret as shown in Figure 14-I. Fill the buret with NaOH and drain it back into the florence flask. Repeat several times until you are sure that the NaOH in the buret is exactly the same concentration as the solution in the florence flask. Protect this solution from contamination. You are going to assume its concentration is unchanged throughout all remaining parts of the experiment.

3. Practice titrating to the end point. Place 20 ml of distilled water in a 250 ml erlenmeyer flask. Add one drop of 1 N HCl and five drops of phenolphthalein. Add NaOH from the buret until one drop causes the clear solution to become light pink. As you add NaOH swirl the flask as shown in Figure 14-II. When the solution is pink (or red) add another drop of HCl and titrate to the end point again. When you are confident that you can detect the approach of the end point, and can hit the end point exactly, proceed to the next part.

4. Clean and dry a 50 ml beaker. Weight it to ± .001 g. Add 0.7 to 0.8 g of potassium acid phthalate. Reweigh to ± 0.001 g. Add about 20 ml of distilled water to the beaker with your squeeze bottle. Pour this solution into a clean 250 ml erlenmeyer flask. Rinse the beaker with the squeeze bottle until *all* of the acid has been transferred to the erlenmeyer flask. Add 5 drops of phenolphthalein.

5. Refill the buret. Read the volume of base to ± .01 ml (even though the marks are at 0.1 ml). Use a piece of white paper behind the buret and read the bottom of the meniscus. Titrate until the

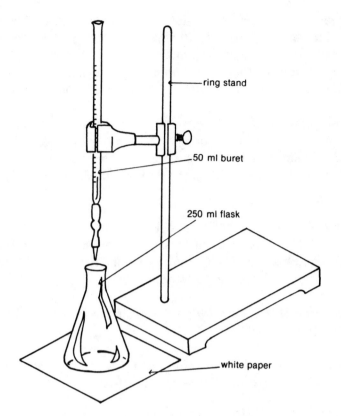

Figure 14-I. Titration set up

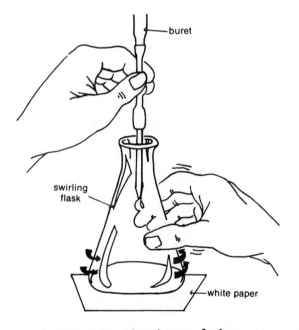

Figure 14-II. Titration technique

pink color will persist for 30 sec or more. If you over shoot the end point the solution is red, you must discard your results. When you have titrated to the end point, read the buret again. Calculate the normality of the base.

89

6. Rinse the erlenmeyer flask with several portions of distilled water. Weigh another potassium acid phthalate sample and transfer it to the erlenmeyer flask. Add five drops of phenolphthalein. Refill the buret. Read the buret, and then titrate as before. Reread the buret, and calculate the normality of the base. This value should be within 1% of the previous value. If it is not, repeat the standardization. Calculate the average normality of the base from all valid experimental values.

7. Prepare an unknown acid solution. Weigh a clean, dry beaker to ±.001 g. Add 1.2 to 1.4 g of unknown acid. Reweigh to ±.001 g. Dissolve the acid in 20 ml of distilled water. Transfer the acid into a clean 100 ml volumetric flask. Rinse the beaker with several 5 ml portions of distilled water to be sure that all the acid is transferred to the flask. Fill the flask to the 100 ml mark and mix the solution well.

8. Rinse a clean pipete with the acid solution. Fill the pipete so that the solution is not diluted, discard the rinse liquid. Pipete 20 or 25 ml of acid solution into a clean 250 ml erlenmeyer flask. Add five drops of phenolphthalein. Fill the buret; read the buret. Titrate to the end point. Reread the buret. Calculate the normality of the acid solution.

9. Titrate additional portions of the acid solution. Calculate the average value of the normality of the acid, using all valid experimental values.

10. Calculate the equivalent weight of the unknown acid. Suggest the acid you tested from the list of possible unknown acids provided by your teaching assistant.

Part A: Get M of NaOH
   1. Get about 200 mL of NaOH
   2. Rinse buret w/ 2 10mL portions (into waste beaker)
   3. Fill buret to 0.00 (somewhere near the top)
   4. Obtain in a flask about 0.5g KHP w/ 2 balance method
   5. Add 25-50 mL of H_2O + Phenolthaliene
   6. Titrate until you see a light pink

Part B: Get GEW of unknown acid A
   1. Get 0.2g of unknow acid
   2. Add H_2O and phenothaline (as in step 5)
   3. Titrate (as in step 6)

Calculation

$$GEW = \frac{\text{weight of acid}}{\text{volume of NaOH} \cdot \text{molarity of NaOH}}$$

$$GEW = \frac{\text{mass unk. A}}{V_{NaOH} \cdot M_{NaOH}}$$

90

Experiment 15
Analysis of a Stomach Antacid Tablet

Object

The amount of acid that can be neutralized by several commercial antacid tablets will be determined.

Background

There are a number of commercial antacid tablets available. Although they contain different chemicals, they all have one property in common: all antacid tablets contain as their primary active component a base. The base present in the antacid tablet reacts with some of the stomach acid to neutralize it. For example milk of magnesia tablets contain $Mg(OH)_2$ and Rolaids contain $NaAl(OH)_2CO_3$.

$$Mg(OH)_2 + 2H^+ \longrightarrow Mg^{2+} + 2H_2O$$

$$NaAl(OH)_2CO_3 + 4H^+ \longrightarrow Na^+ + Al^{3+} + 3H_2O + CO_2$$

The total amount of hydrochloric acid (stomach acid) which can be neutralized by a given antacid tablet is determined by the amount and type of basic material contained in the tablet. The amount of neutralization capacity of an antacid tablet can be determined by titrating a known amount of the substance with a standard solution of hydrochloric acid using a suitable indicator to detect the end point. The indicator methyl orange will be used in this experiment. Methyl orange in an appropriate indicator for titrating a weak base (such as is in antacid tablets) with a strong acid. At the end point of the titration the indicator changes from yellow-orange to red.

The moles of HCl consumed is a direct measure of the strength of the tablet. The number of *milli*moles of HCl consumed simply equals the molarity of the HCl solution multiplied by the volume of HCl used in the titration (in milliliters).

$$\text{mmoles HCl} = (\text{molarity of HCl}) \times (\text{ml of HCl})$$

The moles of HCl consumed equals the number of moles of base in the antacid tablet.

Experimental Procedure

1. Record the molarity of the standard hydrochloric acid that is furnished for this experiment. Clean a buret, and rinse it with two 5-ml portions of the standard acid. Mount the buret on a buret stand, making sure that it is level. Fill the buret with the standard acid to slightly above the zero mark. Make certain that there are no air bubbles in the tip, and run enough acid out so that the meniscus is at the zero mark or below, wiping off the tip of the buret. Record the buret reading.

2. Place one of the antacid tablets furnished by the instructor into a small beaker, add 100 ml of distilled water to it, and crush it with a glass stirring rod to get a solution or slurry. Transfer the solu-

tion in the beaker into a 250-ml conical flask, washing any residue from the beaker to the flask with a fine stream of distilled water from your wash bottle.

3. Add 10 drops of methyl orange indicator to the flask, and then begin to add acid from the buret slowly to the flask. Place a sheet of white paper under the flask so you can observe the color changes more easily. The initial portion of hydrochloric acid may make the solution turn pink. Swirl the solution in the flask until its color changes back to yellow-orange. As the titration proceeds reduce the volume of hydrochloric acid you add each time. As you approach the end point each drop of acid will turn the solution red-orange, which then vanishes as you swirl the solution. A red color which persists for about one minute indicates the end point. Record the buret reading.

4. Repeat the procedure with a second antacid tablet.

Name_____   Lab Section_____

Section_____   Date_____

Analysis of a Stomach Antacid Tablet

Molarity of standard HCl   _____

	Sample #1	Sample #2
Initial buret reading	_____ml	_____ml
Final buret reading	_____ml	_____ml
Volume of acid used	_____ml	_____ml
Millimoles of HCl consumed*	_____	_____
Moles of HCl consumed	_____	_____
Moles of weak base per tablet	_____	_____
Average moles of weak base per tablet		_____

*Show your calculations for Sample #1

What are the active ingredients listed on the container from which you obtained your tablet?

If you overshot the endpoint during the titration, is there any way to "back titrate" as is done when titrating HCl with NaOH solution? Explain.

Experiment 16
Acid-Base Equilibria

Object

In today's experiment you will use an electronic pH meter to investigate weak acid, and buffer solutions, and to measure a titration curve for an acid-base reaction.

Background

pH is defined as $-\log[H^+]$, that is the negative logarithm of the molar concentration of hydrogen ion. It is a measure of the free hydrogen ion in the solution. Experimental measurement of pH are made with an electronic meter fitted with two electrodes. One electrode, called a glass electrode, is sensitive to pH; the other electrode is a reference electrode. Recently, the glass and reference electrodes are available in a single glass tube, called a combination glass electrode.

You will use a pH meter and combination glass electrode to measure the pH of several solutions containing a known concentration of a weak acid. The nature of the acid is unknown. You should calculate the ionization constant of the acid. The ionization reaction may be written $HA \rightleftharpoons H^+ + A^-$ and the constant for the reaction is

$$K_i = \frac{[H^+][A^-]}{[HA]}$$

In a solution containing only the weak acid and water, you may assume all H^+ came from the dissociation of the acid and that $[H^+] = [A^-] = 10^{-pH}$ This enables you to calculate K_i for the unknown acid in solutions A, B and C.

You will also calculate K_i for the unknown acid in solution D, E and F. These solutions will be made by mixing the unknown acid and the sodium salt of the unknown acid. This mixture is called a buffer and will tend to resist a change in pH. You will mix the unknown acid solution, HA, and the sodium salt solution, NaA, to form the buffer solution. You will measure the pH of the solution and then calculate $[H^+]$, $[A^-]$ and $[HA]$ in the solution. These concentrations will be used to calculate K_i for the unknown acid in each solution. The concentration of H^+ is calculated from the pH. The concentration of A^- is calculated from the dilution of the NaA solution. NaA is completely dissociated into Na^+ and A^- ions, and can be considered as the only source of A^- in the buffer solutions. The concentration of HA is calculated from the dilution of the HA solution. In solution D, for example, 20 ml of HA solution is diluted to a total volume of 22 ml. You may assume that essentially all the HA in the buffer solution is due to HA in the original unknown acid solution.

You will next study buffer solutions to see that they do resist changes in pH. You will measure the pH change caused by adding 2 ml of .01 M HCl to a buffer solution. You will compare that to the change caused by adding 2 ml of .01 M HCl to the same quantity of water. You will then repeat the measurements using 0.01 M NaOH.

Finally you will measure a titration curve for the unknown acid. You will place a 25 ml sample of the acid in a beaker, and measure pH as NaOH solution is added. You will graph pH vs. ml of base. The graph will be used to estimate the end point in the titration, and another K_i value for the unknown acid.

During the titration of HA with NaOH, you may assume that the reaction is HA + NaOH $\longrightarrow$ NaA + H_2O. Each mole of NaOH added will produce one mole of A^- in the solution and one mole of HA will be consumed. The actual calculation of K_i is similar to the calculation in a buffer solution. Remember that the total volume of the solution is equal to the initial volume of the HA solution (25 ml) plus the volume of NaOH added.

A summary of the solutions to be studied is listed below.

Soln	Components
A	20 ml HA (stock solution, not diluted)
B	10 ml HA + 10 ml water
C	10 ml HA + 90 ml water
D	20 ml HA + 2 ml NaA
E	10 ml HA + 10 ml NaA
F	2 ml Ha + 20 ml NaA
G	25 ml HA + 25 ml NaA

These solutions require 97 ml HA and 57 ml NaA.

The titration will require 25 ml of HA and about 50 ml of NaOH.

Procedure

The pH meters with combination glass electrodes attached will be available in your laboratory. A small group of students should work with each meter. Directions below are for the measurements to be performed by each group.

The combination glass electrode should not be kept out of water for more than five minutes. The bulb at the tip of the electrode is very fragile. Be careful not to bump or scratch the bulb. Check the pH meter-electrode combination first. Rinse the electrode with a stream of distilled water from your squeeze bottle and let the electrode drain a minute. Place the electrode in a "pH 7 Buffer" solution. Consult with your Teaching Assistant if the melter does not read pH 7.00. Rinse the electrode with distilled water and allow it to drain for 1 minute before each pH reading. When you are not using the apparatus, leave the electrode immersed in distilled water.

1. Obtain a 2 ml pipet (or a 5 ml transfer pipet), a 10 ml pipet and 25 ml pipet from the stockroom. Clean the pipets with soapy water and rinse well (tap water and then distilled water). Clean and dry several 50 ml beakers.

2. Obtain about 120 ml of the unknown acid solution in a clean, dry 150 ml beaker. Record the exact concentration of the acid solution. Cover the beaker to prevent contamination of this solution. Do not permit this solution to be diluted. (A common error would be permitting water in a wet pipet to mix with the solution. If you stick a pipet into this solution — make sure the outside of the pipet is dry and do not let any of the diluted solution (inside the pipet) get into the solution.)

3. Pour about 20 ml of the solution into a clean, dry beaker; this is Solution A. Rinse the glass electrode and permit it to drain a minute. Read the pH of solution A. Calculate the concentration of H^+, A^- and HA in the solution. Calculate K_i for the unknown acid.

4. Pipet 10 ml of distilled water into a clean, dry 50 ml beaker. Dry the outside of the pipet and rinse it with several 1 ml portions of the acid solution. Discard each rinsing and take care not to

dilute the acid solution. Pipet 10 ml of acid into the 50 ml beaker to form Solution B. Mix the solution well and then measure its pH. Calculate the concentrations of H , A and HA and the value of K_i for the unknown acid.

5. Obtain a 100 ml graduated cylinder, it must be clean and rinsed with distilled water. It need not be dry. Pipet 10 ml of the unknown acid into the graduated cylinder and add distilled water to bring the bottom of the meniscus to the 100 ml mark; this is Solution C. Pour the solution back and forth several times from the cylinder to a clean, dry beaker (100 or 150 ml). When the solution is mixed well; read the pH, calculate the concentrations and K_i, as before.

6. Prepare solutions D, E and F in clean dry 50 ml beakers. Pipet two 10 ml portions of HA into beaker D, pipet 10 ml of HA into beaker E. Rinse the 2 ml pipet (or the 5 ml transfer pipet) with HA solution. Pipet 2 ml of HA solution into beaker F. Obtain about 75 ml of NaA solution in a clean, dry 100 ml beaker. Rinse a clean 2 ml pipet with this solution and pipet 2 ml of NaA solution into beaker D. Rinse a clean 10 ml pipet with the NaA solution. Pipet 10 ml of NaA into beaker E. Pipet two 10 ml portions of NaA solution into beaker F. Measure the pH of solutions D, E and F. Calculate the concentrations of H^+, A^- and HA in each solution. Calculate the value for K_i in each solution.

7. Prepare Solution G by pipeting 25 ml of NaA and 25 ml of HA into a 50 ml beaker. Pour the solution into a clean dry 50 ml beaker, and pour it back to the original beaker. Pour back and forth several times to mix the solution well. Place half of Solution G in one beaker. Read the pH. Pipet 2 ml of 0.01 M HCl into the solution. Mix well and read the pH. Calculate the pH change caused by adding HCl to Solution G. Place 25 ml of distilled water in a clean beaker. Measure the pH and then pipet 2 ml of 0.01 M HCl into the water. Mix well and read the pH. Calculate the pH change caused by adding HCl to water. Repeat this series of measurements to obtain the pH change caused by adding 2 ml of 0.01 M NaOH to 25 ml of Solution G and the pH change caused by adding 2 ml of NaOH to 25 ml water. Comment on the ability of the buffer solution (Solution G) to resist changes in pH.

8. Pipet 25 ml of HA into a clean, dry 50 ml beaker. Add five drops of phenolphthalein to the solution. Carefully place a magnetic stirring bar in the solution. Adjust the electrode so that it clears the stirring bar. The spinning bar will break the tip of the electrode if clearance is not maintained.

9. Obtain 50 ml of 0.15 M NaOH in a beaker. Check out a 50 ml buret. Clean it with soap and water and a buret brush. Rinse with tap water and then with four 5 ml portions of distilled water. Pour the NaOH into the buret. Don't worry if the NaOH solution is diluted by the water clinging to the inner walls of the buret. Drain the contents of the buret back into the beaker which held the NaOH solution. Refill and drain the buret several times until the solution in all parts of the buret is uniform. Drain the buret until the buret reads 0.00 ml.

10. Read the pH of the HA solution. Add NaOH from the buret until about 4 ml have been added. Keep the solution well stirred with the magnetic stirrer. Record the pH of the solution and the exact buret reading. Add more NaOH, and then record pH and buret readings again. Repeat this procedure and try to obtain readings at 0, 4, 8, 12, and 14 ml. Record values at 0.5 ml intervals above 14 ml, until the pH reaches a value of 10. Also record the buret reading at the point where the solution retains a definite pink color for 30 seconds. Perform the calculations outlined on the data sheet.

Name_____ Section_____

Station_____ Date_____

Acid-Base Equilibria

The unknown acid solution is_____ M in HA.

A.   Solution A is_____M in HA. Solution A has a pH of_____.

   $[H^+]$ =_____ M, $[A^-]$ =_____ M, $[HA]$ =_____M.

   Show your calculations and assumptions in this space.

For the reaction $HA \rightleftharpoons H^+ + A^-$, the equilibrium constant is K_i.

In Solution A, the value of K_i =_____. Show your work.

B.   Solution B is_____ml HA stock plus_____ml water. $[HA]$ =_____M.

   Measured pH =_____ $[H^+]$ =_____M; $[A^-]$ =_____M

   The value of K_i, in solution B, is_____. Show your work.

C.   Solution C is_____ ml HA stock plus_____ml water. $[HA]$ =_____M.

   Measured pH_____. $[H^+]$ =_____; $[A^-]$ =_____

   The value of K_i, in solution C, is_____. Show your work.

D. The NaA solution is_____M

   Solution D is_____ml of HA plus_____ml of NaA.

   [HA] =_____M; [A⁻] =_____M

   Measured pH is   Calculated [H⁺] =   M . Show your work.

   The value for K_i =_____ . Show your work.

E. Solution E is_____ ml of HA plus_____ ml of NaA.

   [HA] =_____M; [A⁻] =_____M

   Measured pH =_____; calculated [H⁺] =_____M

   The value for K_i in solution E is_____. Show your work.

F. Solution F is_____ ml of HA plus_____ml of NaA.

   [HA] =_____M; [A⁻] =_____M

   Measured pH =_____ ; Calculated [H⁺] =_____ M

   The value for K_i in solution F is_____. Show your work.

Fill in the table below to summarize the Ki values you have obtained.

Solution	pH	[H⁺]	[A⁻]	[HA]	K_i
A					
B					
C					
D					
E					
F					

Name_____   Section _____

Station_____   Date_____

In the space below, write out your comments on the K_i values you have obtained.

G. (1) The pH of Solution G is_____ ; after addition of 2 ml of 0.01M HC1 to 25 ml of

Solution G, pH is_____ .

The pH change is_____

The pH of pure water is_____; after addition of HC1 the pH is _____.

The pH change is_____.

(2) The pH of solution G is_____ ; after addition of 2 ml of 0.01 M NaOH, the

pH is_____ .

The pH change is_____.

The pH of pure water is_____; after addition of NaOH the pH is_____.

The pH change is_____.

Write out your comments on the ability of Solution G to resist changes in pH.

Titration data for HA plus NaOH.

Solution being titrated is_____ml of _____M HA.

Volume of base	pH	Volume of base	pH
0.00 ml			

103

Prepare a graph of pH (Y-axis) vs. ml of base (X-axis).

Draw a smooth curve through the data points.

The phenophthalein end point is the point where the solution remained pink for 30 sec. This

occurred when_____ml of NaOH had been added.   Calculate the concentration of the
NaOH solution from this data. Show your work.

The "electronic end point" is the steepest part of the pH vs Volume curve. This end point

occurred when_____ ml of NaOH had been added. Which end point is easiest to measure?
Which end point is most accurate?

Calculate the K_i for the unknown solution at the point where 6.00 ml of NaOH has been added to

the acid. The total volume of solution is_____ ml; the pH is_____; $[H^+]$ =_ _____;

$[A^-]$ =_____; $[HA]$ =_____

Show your work below.

Experiment 17
Measurement of pH

Object

In this experiment you will use an electronic pH meter to measure the pH of several aqueous solutions and to measure the pH change produced by diluting and adding acid to these solutions.

Background

The pH scale measures the acid content of solutions. It is most often used with aqueous solutions (i.e., solutions made with water as the solvent), where values generally run from 0 to 14. pH is defined by the equation: $pH = -\log [H^+]$ or pH is the negative logarithm of the concentration of hydrogen ion. In 1 M HCl (made by dissolving 1 mole of HCl in enough water to produce 1 liter of solution) the H^+ concentration will be one mole per liter or $[H^+] = 1$ M. The log of 1 is 0 since $10^0 = 1$ (ten raised to the zeroth power is one). The table below shows some other solutions, the concentrations of H^+ and OH^- ions which will be present and the pH. *Note the following facts about pH values of solutions:

(a) Changing $[H^+]$ by a factor of 100, changes pH by 2 since $\log 100 = 2$ or $\log .01 = -2$.

(b) Changing $[H^+]$ by a factor of 2, changes pH by 0.3 since $\log 2 = 0.3$.

(c) Diluting any solution will move it closer to pH 7.

(d) In any solution $[H^+] [OH^-] = 10^{-14}$. In a neutral solution, $[H^+] = [OH^-] = 10^{-7}$.

(e) In 1 M HCl, $[H^+] = 1$ M and in 0.01 M HCl, $[H^+] = 0.01$ M or one mole of HCl produces one mole of H^+.

(f) In 1 M NaOH, $[OH^-] = 1$ M and in 0.01 M NaOH, $[OH^-] = 0.01$ M or one mole of NaOH produces one mole of OH^-.

(g) Acidic solutions have pH values less than 7, a neutral solution has a pH of 7; and basic solutions have pH values greater than 7.

SOLUTION	$[H^+]$	$[OH^-]$	$\log[H^+]$	pH
acid, 1 M HCl	1 M	10^{-14}M	-0.0	0.0
dilute acid, 0.02 MHCl	.02	5×10^{-13}	-1.7	1.7
very dilute acid, 0.01 M HCl	.01	10^{-12}	-2.0	2.0
normal distilled water, some CO_2	10^{-5}	10^{-9}	-5.0	5.0
very pure distilled water	10^{-7}	10^{-7}	-7.0	7.0
very dilute base, 0.01 M NaOH	10^{-12}	.01	-12.0	12.0
dilute base, 0.02 M NaOH	5×10^{-13}	.02	-12.3	12.3
base, 1 M NaOH	10^{-14}	1	-14.0	14.0

Strong acids such as HCl and HNO_3 are completely dissociated into ions. One mole of HCl or HNO_3 will produce one mole of H^+ ions and one mole of negative ions (Cl^- or NO_3^-) in solution. A 0.01 M HCl solution should give $[H^+] = 0.01$ or a pH of 2.0. Diluting a strong acid solution by a factor of two should raise the pH by 0.3 since the dilution should lower the concentration of H^+ by a factor of two.

Weak acids, such as acetic acid or formic acid, are only partially dissociated into ions. When one mole of acetic acid is dissolved in water, much less than one mole of H^+ will be produced in the solution. The pH of a 0.01 M acetic acid solution will be greater than two. Diluting an acetic acid solution should increase the pH less than 0.3 units. According to LeChatelier's Principle, the stress of diluting the solution will cause a larger fraction of the acetic acid to dissociate into ions. Diluting the solution by a factor of two, should decrease the concentration of H^+ by a factor less than two and therefore should change the pH less than 0.3.

In a similar way, bases can be strong or 100% dissociated into ions in solution; NaOH is a strong base. Ammonium hydroxide, NH_4OH, is a weak base and is only partially dissociated into ions.

Salts formed by mixing strong acids (HCl or HNO_3) with strong bases (NaOH) should give solutions with a pH of 7. Salts formed from a strong acid and a weak base such as ammonium chloride, will give a pH less than 7. On the acid side you might say, because the acid overpowers the base. Salts formed with a weak acid and a strong base, such as sodium acetate, will give a pH greater than seven.

Buffer solutions are special solutions which maintain a fairly constant pH in spite of chemical stresses. Blood and wine are examples of highly buffered solutions. Some buffers contain a weak acid and the salt of a weak acid in solution. This two component mixture is quite resistant to changes in pH, even if small amounts of acid or base are added to the solution. Wine and solutions made with acetic acid + sodium acetate are examples of these buffers.

Experimental measurement of pH is done today using an electronic instrument fitted with two electrodes. Recently both electrodes have been placed inside the same glass tube. The active part of the electrode contains a fragile bulb made of special glass. Periodically the operation of the pH meter-glass electrode instrument is checked by observing the readings produced by buffer solutions of known pH.

Experimental Procedure

1. Perform this experiment in groups. Each group should choose a pH meter. Observe the following rules regarding the glass electrode:

    (a) Be very careful with the tip of the electrode. It is very thin glass and therefore is easily broken.

    (b) Do not allow the glass electrode to dry out more than one minute. When not in use it should be immersed in distilled water.

    (c) Before each reading, squirt off the electrode with a small portion of distilled water. Be careful to rinse the entire surface of the electrode.

2. Turn the pH meter on. Obtain 20 ml of "Buffer-pH 7.00" in a clean, dry 50 ml beaker. Rinse the glass electrode with distilled water from your squeeze bottle. Collect the rinse water in a 250 ml beaker. Immerse the electrode in the buffer solution. If the reading is not pH 7.00, consult your

TA for further instructions. Your pH meter should now read accurate pH values for a week or more if the glass electrode is properly cared for. Discard the buffer solution.

3. Measure 20.0 ml samples of each of the following solutions in clean, dry graduated cylinders. Place the sample in a clean, 50 ml beaker, stir, and read the pH of the sample after twofold dilution. Finally, add one drop of 0.01 m HCl, stir the solution and read the pH of the final mixture. Rinse the electrode with distilled water between samples. Use the following samples: 0.01 M HCl, 0.01 M NaCl, 0.01 M acetic acid, 0.01 M NH_4OH, 0.01 M NaOH, distilled water, city water.

4. Measure the pH of a 20 ml sample of 0.01 M NH_4Cl. Add 20 ml of 0.01 M NH_4OH, mix well and record the pH. Transfer the solution to a larger beaker and add 40 ml of distilled water; mix well and record the pH. Add one drop of HCl, mix and record the pH.

5. Repeat the pH measurements using 20 ml of 0.01 M $NaC_2H_3O_2$ first. Then add 20 ml of 0.01 M $HC_2H_3O_2$ (acetic acid), read the pH; add 40 ml of distilled water and read the pH. Add one drop of HCl and read the pH.

Name _____   Section _____

Station _____   Date _____

Measurement of pH

Step 3

SAMPLE	pH of ORIGINAL SAMPLE	pH AFTER TWO FOLD DILUTION	pH AFTER ONE DROP HC1
0.01 M HCl	_____	_____	_____
0.01M NaCl	_____	_____	_____
0.01 M acetic acid	_____	_____	_____
0.01 M NH_4OH	_____	_____	_____
0.01 M NaOH	_____	_____	_____
Distilled Water	_____	_____	_____
City Water	_____	_____	_____

Step 4

pH of 0.1 M NH_4Cl _____

pH of 40 ml of HH_4C1-NH_4OH_____

pH of 80 ml of NH_4C1-NH_4OH_____

pH of mixture + one drop HCl_____

Step 5

pH of 0.01 M $NaC_2H_3O_2$_____

pH of 40 ml of $NaC_2H_3O_2$-$HC_2H_3O_2$_____

pH of 80 ml of $NaC_2H_3O_2$-$HC_2H_3O_2$_____

pH of mixture + one drop HCl _____

CONCLUSIONS BASED ON YOUR pH READINGS

A. From your pH data, list the name and formula of substances (used in this experiment) in the following categories:

111

Strong acid:

weak acid:

strong base:

weak base:

salt which gives neutral solution:

salt which gives acidic solution:

B. Briefly explain the pH changes produced by two fold dilution of HCl, NaOH, $HC_2H_3O_2$ and NH_4OH.

Which should show an increase of 0.3?

Which should show a decrease of 0.3?

C. (a) List solutions which gave a large change in pH from addition of one drop of 0.01 M HCl. Explain each one.

(b) List solutions which gave a moderate change in pH from the addition of HCl. Explain each one.

(c) List solutions which gave a small change in pH from the addition of HCl. Explain each one.

Experiment 18
The Equilibrium Constant of a Complex

Object

In this experiment you will make several determinations for the value of the equilibrium constant that describes the formation of a $FeSCN^{2+}$ from Fe^{3+} and SCN^-. You will quantitatively prepare several solutions and then measure their absorbance using a spectrophotometer. The absorbance gives the concentration of $FeSCN^{2+}$. The concentrations of Fe^{3+} and SCN^- are obtained as the difference between the initial concentration and the concentration consumed by the formation of the $FeSCN^{2+}$

Background

The blood-red complex, $FeSCN^{2+}$, is formed when SCN^- solutions and Fe^{3+} solutions are mixed. The reaction is:

$$Fe^{3+} + SCN^- \rightleftharpoons FeSCN^{2+}$$

which has an equilibrium constant, K, given by:

$$K = \frac{[FeSCN^{2+}]}{[Fe^{3+}]\ [SCN^-]}$$

In today's experiment, you will accurately dilute stock solutions of Fe^{3+} and SCN^- to give a series of solutions containing certain initial concentration values. You will measure the absorbance (color intensity) of each solution. Since Fe^{3+} and SCN^- are colorless, the absorbance of the solution can be used to calculate the concentration of $FeSCN^{2+}$ in the solution. The final concentration values of Fe^{3+}, SCN^-, and $FeSCN^{2+}$ are inserted into the equilibrium constant expression to obtain an experimental value for K. Theory tells us that K will be constant; you will determine the range of K values which you obtain from your experimental data.

You will utilize a Spectronic 20 spectrophotometer for measuring the absorbance of the solutions. You will first adjust the zero setting on the instrument; no light passes through the machine during this test. This corresponds to a deeply colored solution with a high absorbance. Next the light control is adjusted to read zero absorbance when a clear solution (pure water) is placed in the instrument. All readings should be made at a wavelength of 450 mμ (blue light). Notice that this complex looks red when it absorbs blue light. At this wavelength, the absorbance by the $FeSCN^{2+}$ complex is at the maximum value. The Fe^{3+} and SCN^- do not absorb at this wavelength. There is a direct relationship between the absorbance of the solution and the concentration of the complex; this relation is described by the equation:

$$A = k\ [FeSCN^{2+}]$$

You will deal with nine different solutions during this experiment. Be very careful to use proper volumetric procedures when mixing each solution. Review that section of the Introduction if necessary.

113

Good results can be obtained only if the solutions are carefully prepared so that the dilution factors are accurate. Solutions A-D will contain 10 mL of KSCN stock solution and 10 mL of an Fe^{3+} solution. These solutions are prepared according to the Table below:

Solution	SCN^-	Fe^{3+}
1	—	10 mL stock $Fe(NO_3)_3$ + 15 mL H_2O
2	—	10 mL soln 1 + 15 mL H_2O
3	—	10 mL soln 2 + 15 mL H_2O
A	10 mL stock KSCN	10 mL stock $Fe(NO_3)_3$
B	10 mL stock KSCN	10 mL soln 1
C	10 mL stock KSCN	10 mL soln 2
D	10 mL stock KSCN	10 mL soln 3

Recall that dilution of a solution is described by the equations:

$$M_1 V_1 = M_2 V_2 \qquad \text{or} \qquad M_2 = M_1 (V_1/V_2)$$

The (V_1/V_2) term is the dilution factor. Let us illustrate the use of this equation in obtaining the concentrations of SCN^- and Fe^{3+} in Soln A. This solution contains 10 mL of stock KSCN solution, 2×10^{-4} M, so $V_1 = 10$ mL and $M_1 = 2 \times 10^{-4}$ M. It is diluted by the addition of 10 mL of Fe^{3+} solution so $V_2 = 20$ mL, the total volume. In Soln A we calculate the concentration of SCN^- as:

$$M_2 = M_1 (V_1/V_2) = 2 \times 10^{-4} (10/20) = 1 \times 10^{-4} \text{ M} = [SCN^-]$$

The stock Fe^{3+} solution is 5×10^{-2} M, so in Soln A we calculate the concentration of Fe^{3+} as:

$$M_2 = 5 \times 10^{-2} (10/20) = 2.5 \times 10^{-2} \text{ M} = [Fe^{3+}]$$

The concentrations of Fe^{3+} in Solns 1-3 can be calculated by a similar procedure. Soln 1 contains 10 mL of stock $Fe(NO_3)_3$ which is diluted to 25 mL total volume. Soln 2 contains 10 mL of Soln 1 which has been diluted to 25 mL total volume. Soln 3 contains 10 mL of Soln 2 which is diluted to 25 mL total volume.

In Soln A the Fe^{3+} concentration is 250 times greater than the SCN^- concentration (2.5×10^{-2} M/1×10^{-4} M = 250). Under these conditions, it is correct to assume that all the SCN^- will be converted into the complex. This means that after the Fe^{3+} and SCN^- react in Soln A that the concentration of $FeSCN^{2+}$ is 1×10^{-4} M. From the absorbance measured for Soln A, and this concentration of $FeSCN^{2+}$, you can obtain a value for k. This value of k can be used to calculate the concentrations of $FeSCN^{2+}$ in Solns B-D from the measured absorbance values.

The experimental value of the equilibrium constant, K, for Solns B-D will be calculated from the experimental concentrations of $FeSCN^{2+}$, unreacted Fe^{3+} and unreacted SCN^-. The concentration of $FeSCN^{2+}$ is obtained from the absorbance reading. The concentration of unreacted Fe^{3+} (or Fe^{3+} which remains after the reaction of Fe^{3+} and SCN^-) is calculated as the difference between the original Fe^{3+} concentration and the Fe^{3+} used to form the $FeSCN^{2+}$. The concentration of unreacted SCN^- is calculated as the difference between the original SCN^- concentration and the SCN^- used to form the $FeSCN^{2+}$. To

$KSCN = 2 \times 10^{-4}$ M

illustrate these calculations we will describe a hypothetical solution. Assume KSCN and $Fe(NO_3)_3$ were mixed so that the concentrations are $[SCN^-] = 1 \times 10^{-4}$ M and $[Fe^{3+}] = 3 \times 10^{-3}$ M. These numbers represent the concentrations after dilution, but assume no reaction has occurred to produce $FeSCN^{2+}$. Assume that the absorbance measurement gives us a concentration of the complex ion of $[FeSCN^{2+}] = 2.5 \times 10^{-5}$ M. We can calculate the concentration of unreacted Fe^{3+} as: $[Fe^{3+}] = (3 \times 10^{-3}$ M$) - (2.5 \times 10^{-5}$ M$) \approx 3 \times 10^{-3}$ M and the concentration of unreacted SCN^- as:

$$[SCN^-] = (1 \times 10^{-4} \text{ M}) - (2.5 \times 10^{-5} \text{ M}) = 7.5 \times 10^{-5} \text{ M}.$$

With more Fe^{3+} than SCN^- in the mixture, it can be seen that essentially all the Fe^{3+} remains unreacted, but only a portion of the SCN^- remains unreacted.

Procedure

Perform this experiment with a lab partner.

1. Check out a 10 ml pipet and a 25 ml volumetric flask from the stock room. Clean them with soap and water; rinse with tap water and distilled water. This glassware need *not* be dried.

2. Obtain four clean dry 50 ml Erlenmeyer flasks. Label them (with pencil) A, B, C, D. Obtain about 50 ml of stock KCNS solution, 2×10^{-4} M. Be sure that you only pour this solution into a clean, dry container. Carefully rinse the 10 ml pipet with several 2 ml portions of the KCNS solution and discard each solution after you use it to wet all inner parts of the pipet. Remember to use a safety bulb to fill the pipet.

3. When the pipet has been rinsed thoroughly with the KCNS solution, carefully pipet 10 ml portions of the 2×10^{-4} M KCNS solution into each of the 50 ml Erlenmeyer flasks (A,B,C and D). Rinse the pipet with distilled water.

4. Rinse the pipet with several 2 ml portions of the stock $Fe(NO_3)_3$ solution, 5×10^{-2} M. Pipet 10 ml of the 5×10^{-2} M $Fe(NO_3)_3$ solution into flask A.

5. Pipet 10 ml of 5×10^{-2} M $Fe(NO_3)_3$ solution into the 25 ml volumetric flask. Carefully add 5 ml of 2.5 M HNO_3 (graduated cylinder is accurate enough) to the flask. Add distilled water until the total volume of the solution is exactly 25 ml. The bottom of the meniscus should be up to the mark on the volumetric flask. Place the glass or plastic stopper on the volumetric flask and invert the flask many times to mix the solution thoroughly. Pour this solution into a clear, dry container and label it solution 1.

6. Rinse the volumetric flask with four 5 ml portions of distilled water. Rinse the pipet with three 2 ml portions of distilled water. Rinse the pipet with three 2 ml portions of Solution 1. Pipet 10 ml of Solution 1 into flask B. Pipet another 10 ml of Solution 1 into the volumetric flask.

7. Add 5 ml of 2.5 M HNO_3 to the 10 ml of Solution 1 in the volumetric flask. Dilute to 25 ml. Mix thoroughly and pour this solution, Solution 2, into a clean, dry container.

8. Rinse the volumetric flask and pipet as before. Pipet 10 ml of solution 2 into flask C. Pipet 10 ml of solution 2 into the volumetric flask, add 5 ml of 2.5 M HNO_3 and dilute to 25 ml.

115

9. Mix the contents of the volumetric flask well; this is Solution 3. Rinse the pipet with distilled water and then with Solution 3. Pipet 10 ml of Solution 3 into flask D.

10. Check your solutions at this time to make sure that the components you mixed correspond to those shown in the Tables presented in the Background section.

11. Take Solutions A, B, C and D to a Spectronic 20. Obtain an absorbance reading for each solution at 450 mμ. Refer to pages 4 and 5 for operating instructions of the Spectronic 20 instrument. Be sure to line up the special test tube correctly during your calibration and during each reading. Be sure to rinse the test tube well with each new solution to be measured.

Experiment 19
Spectrophotometric Chromium Analysis

Object

In today's experiment, you will perform a quantitative analysis for chromium in an unknown salt. The chromium salt is converted into Cr(VI), chromate, and a spectrophotometric method is used to measure the chromium concentration.

Background

An unknown chromium salt will be issued to you. Possible salts include $Cr_2(SO_4)_3$, $CrCl_3 \cdot 6H_2O$, $K[Cr(SO_4)_2] \cdot 12H_2O$, or $CrPO_4 \cdot 2H_2O$. Each of these salts contain Cr(III), chromium in the +3 oxidation state. Many different forms of Cr(III) are possible in solution; each has a characteristic color and a certain intensity of color.

For quantitative analysis of chromium salts, the salt is converted into Cr(VI) and placed in basic solution. The chromium will be present as chromate, CrO_4^{2-}, regardless of the nature of the Cr(III) salt. Chromate has a characteristic yellow color and will show maximum absorbance of light with a wavelength of 373 nm. You will measure the intensity of absorbtion of light, or absorbance, in the chromium solution. This can be used to determine the chromium content.

In most solutions, a plot of absorbance versus concentration will give a straight line. This relationship, known as *Beer's Law*, can be expressed as

$$A_\lambda = E_\lambda lC = kC$$

where A_λ represents the absorbance of the solution at wavelength λ, l is the path length of light through the solution, and C is the concentration of the species which absorbs light. The symbol E_λ is the molar absorbtivity or extinction coefficient. The maximum extinction coefficient for chromate occurs at 373 nm, where $\epsilon_{373} = 4815$ M^{-1} cm^{-1}. In today's experiment, E_λ and l will be constant, so they can by combined in a constant, k. A graph of concentration (x-axis) versus absorbance (y-axis) should be linear with a slope of k. This graph is known as a Calibration Curve. It is prepared from absorbance measurements on a series of solutions of known chromium concentration.

The oxidizing agent which oxidizes Cr(III) to Cr(VI) is hydrogen peroxide, H_2O_2. In basic solution, many Cr(III) salts will give a precipitate of $Cr(OH)_3$. In general, Cr(III) complexes will interconvert into other Cr(III) complexes very slowly. Alkaline peroxide solutions, however, will convert any Cr(III) compound into chromate. An excess of peroxide can be used in this experiment, since any excess can be easily destroyed by boiling the solution. At high temperatures, peroxide will decompose into water and oxygen by the reaction:

$$2H_2O_2 \longrightarrow 2H_2O + O_2$$

121

Procedure

Part 1. Prepare a solution of chromate from your unknown Cr(III) salt. Accurately weigh 0.1 g of the unknown salt into a 100 ml erlenmeyer falsk. Add about 20 ml distilled water and about 10 ml of 2 M NaOH. Heat to about 60°C and add about 10 ml of 15% H_2O_2 to the solution slowly. *Caution = 15% H_2O_2 is EXTREMELY caustic — avoid contact with your hands.* If the solution is not a clear, bright yellow color, add H_2O_2 dropwise until all off-colors are gone. When all Cr(III) has been converted into CrO_4^{2-}, gently boil the solution 5-10 minutes to decompose all excess hydrogen peroxide. Be careful that none of your solution splashes out of the flask during any part of the experiment. Cool the solution to room temperature, and quantitatively transfer it to a 100 ml volumetric flask. Be sure to rinse the erlenmeyer flask with many small portions of distilled water to transfer the chromium. Be sure to rinse the funnel well so that no chromium is present in the solution which adheres to the funnel. Add water so that the bottom of the meniscus is at the 100 ml mark. Label this Solution A. Mix the solution well.

Clean a 5 ml pipete. Rinse it with distilled water and several small portions of Solution A. Pipete 5 ml of Solution A into another, clean 100 ml volumetric flask. Dilute to the 100 ml mark and label this Solution B. Mix the solution well.

Part 2. Use the Spectronic 20 spectrophotometric to measure absorbance. For detailed operating instructions, refer to the Introduction. Turn the Spec. 20 on; select a wavelength of 373 nm. Adjust the Zero Setting and the Light Control. Be sure to align the special test tube correctly for each reading.

Rinse the test tube with several small portions of the test solution before filling the tube. When you have obtained the absorbance reading, discard the solution.

Read absorbance values for several standard solutions of chromate in the range 2×10^{-5} M to 2×10^{-4} M. Record the concentration and absorbance for each solution. Prepare a Calibration Curve.

Read the absorbance of Solution B. From the Calibration Curve, read the concentration of chromium in Solution B. Calculate the concentration of chromium in Solution A. From the concentration of chromium in Solution A and the composition of Solution A, calculate the per cent chromium in the unknown salt. Calculate the per cent chromium in each of the possible Cr(III) salts. Indicate your choice.

122

Name_____   Section_____

Station_____   Date_____

Spectrophotometric Chromium Analysis

Part 1.   Weight of Erlenmeyer flask_____ g; weight of flask and unknown salt,_____ g.

Weight of unknown Chromium salt in Solution A,_____ g. Describe the unknown salt.

Solution A: _____ g salt in _____ ml.

Solution B: _____ ml of Solution A in_____ ml.

Part 2.   Concentration of Standard   Absorbance at
           Chromate Solution   373 nm

_____ M   _____

_____ M   _____

_____ M   _____

_____ M   _____

_____ M   _____

_____ M   _____

_____ M   _____

Absorbance of Solution B,_____ .

Slope of Calibration Curve,_____ . (Graph is on the back of this sheet).

Concentration of chromium in Solution B,_____ M

Concentration of chromium in Solution A,_____ M

Percent chromium in unknown salt,_____ % Cr. Show your work.

123

The salt which comes closest to this per cent chromium is _____.

$Cr_2(SO_4)_3$ is _____ % Cr.   Show your work.

$CrCl_3 \cdot 6H_2O$ is _____ % Cr.   Show your work.

$K[Cr(SO_4)_2] \cdot 12H_2O$ is _____ % Cr.   Show your work.

$CrPO_4 \cdot 2H_2O$ is _____ % Cr.   Show your work.

Experiment 20
Spectrophotometric Determination of Trace Mercury

Object

Trace amounts of mercury will be extracted from water, and the concentration of the extracted mercury will be determined by using a spectrophotometer.

Background

The heavy metal mercury has received widespread attention as an environmental pollutant. Mercury and mercury compounds are important industrial chemicals, and for many years, industrial wastes containing mercury were dumped into our streams and lakes. In many cases, this mercury eventually ends up in fish via the food chain. Mercury is quite toxic, and in some areas the mercury content of fish is sufficiently high that eating the contaminated fish is dangerous to humans.

The analysis of ppm (parts per million) levels of mercury in silts, clays, waters, as well as fish, requires special techniques. Mercury and many other elements react with organic complexing agents to form intensely colored complexes. The intensity of the color is directly proportional to the amount of mercury (in the complexed form) in solution. Thus, a solution that is "twice as colored" as another contains twice the amount of mercury complex. The intensity of color is easily measured by a spectrophotometer. Specifically mercury reacts with dithizone to produce a orange-colored complex. By comparing the intensity of the color of an unknown sample with the intensity of the color of a known solution with a spectrophotometer, the concentration of the unknown can be determined.

The mercury samples for which we wish to determine the concentration are in water solution. It is necessary to remove the mercury from the water by a technique called solvent extraction. Very simply stated solvent extraction involves shaking two immiscible (non-mixing) liquids together, and a species that is dissolved originally in one liquid will be transferred to the other liquid if that compound is more soluble in the second liquid. Specifically in this case, the mercury that is originally present in an acidic water solution is Hg^{2+} ions can be extracted by a chloroform solution containing dithizone. As the water and chloroform (which are not soluble with each other) are shaken, the Hg^{2+} in the water reacts with the dithizone in the chloroform, forming a mercury·dithizone complex (mercuric dithizonate). This complex is more soluble in the chloroform than in the water, thus the mercury is transferred into the chloroform. The chloroform layer can be removed from the water layer, and the intensity of color due to the mercury-dithizone complex can be measured spectrophotometrically to determine the amount of mercury in the original solution.

The theory and operation of spectrophotometers is discussed in a section at the first of this manual. In this experiment a set of calibration standards will be measured at 500 nanometers (nm). Three samples of unknown solution will be prepared in the same manner as the standards and the absorbances measured at the same wavelength with the spectrophotometer. The mercury content of the unknowns can then be determined from a calibration curve that is prepared by plotting the mercury concentration versus absorbance for the standard solutions.

Experimental Procedure

1. Take 75 ml of standard mercury solution (0.5 microg per ml or 0.5 ppm). Also take a few ml each of nitric acid (HNO_3), acetic acid ($HC_2H_3O_2$), and hydroxylamine hydrochloride. Arrange five clean 150 ml beakers on your desk and label them "blank", 0.25, 0.5, 0.75, and 1.0. With a 100 ml graduated cylinder add to the beakers the quantities of distilled water indicated in the table below. Use a pipette to add to each beaker the volume of standard mercury solution indicated. Finally, add the nitric acid, acetic acid, and hydroxylamine hydrochloride as indicated. Set these beakers aside.

ppm Mercury	ml distilled H_2O	ml standard mercury solution	ml nitric acid	ml acetic acid	ml hydroxylamine hydrochloride
blank	95	0	2	1	1
0.25	90	5	2	1	1
0.50	85	10	2	1	1
0.75	80	15	2	1	1
1.00	75	20	2	1	1

2. Arrange three clean 150 ml beakers on your desk. To two of the beakers add 10 ml of *unknown* mercury solution and to the other beaker add 10 ml of tap water. Then add 85 ml of distilled water, 2 ml nitric acid, 1 ml acetic acid, and 1 ml hydroxylamine hydrochloride to each beaker.

3. Label eight clean test tubes "blank", 0.25, 0.50, 0.75, 1.0, unknown I, unknown II, and "tap water". Transfer the "blank standard" to a 250-ml separatory funnel and add exactly 10 ml of the dithizone chloroform solution: Shake the mixture for 2 min. The proper method for the use of a separatory funnel is illustrated in Figure 20-I. After shaking for 2 min., open the stopcock with the separatory funnel inverted to relieve the pressure. Close the stopcock, turn the separatory upright, and allow the two layers to separate. Open the stopcock and drain the lower layer into the test tube labeled "blank". Discard the water layer.

4. Rinse the separatory funnel with hot water, and twice with distilled water. Repeat the process with each of the other seven samples.

5. To operate the Spectronic 20 spectrophotometer:

    a. Turn on the instrument and allow it to warm up for at least 30 minutes. The instrument should be turned on at the beginning of the laboratory period.

    b. Set the spectrophotometer wavelength control to 500 nm.

    c. Using the zero control knob, adjust the indicator needle on the left side of the scale to zero. The sample holder is empty and the light box is closed.

    d. Transfer enough of the blank solution to fill the sample tube about half full and insert it into the sample holder.

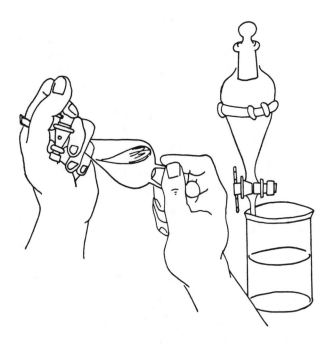

Figure 20-1. Use of the separatory funnel

e. Adjust the indicator needle on the right side of the scale until the meter reads 100 (% transmittance). Absorbance (red scale) should read zero.

f. Fill another sample tube about half full of the 0.25-microgram standard, insert it in the sample holder and read the absorbance. Record the value on the data sheet.

g. Rinse the sample tube with a small portion of the 0.50-microgram standard and then pour it out. Transfer the remainder of the 0.50-microgram standard to the tube and read its absorbance from the dial. Record the value on the data sheet.

h. Repeat these steps with the other standards and the unknowns.

i. Plot absorbance versus concentration for the blank and four standards. Draw a straight line through the data points.

j. Estimate the concentration for each unknown from the calibration curve and report in the space provided on the data sheet.

Name________________________ Lab Section________________

Station______________________ Date____________________

Spectrophotometric – Determination of Trace Mercury

Sample (ppm)	Absorbance	ppm Mercury
blank	0	
0.25	______	
0.50	______	
0.75	______	
1.0	______	
Unknown I	______	______
Unknown II	______	______
Tap water	______	______

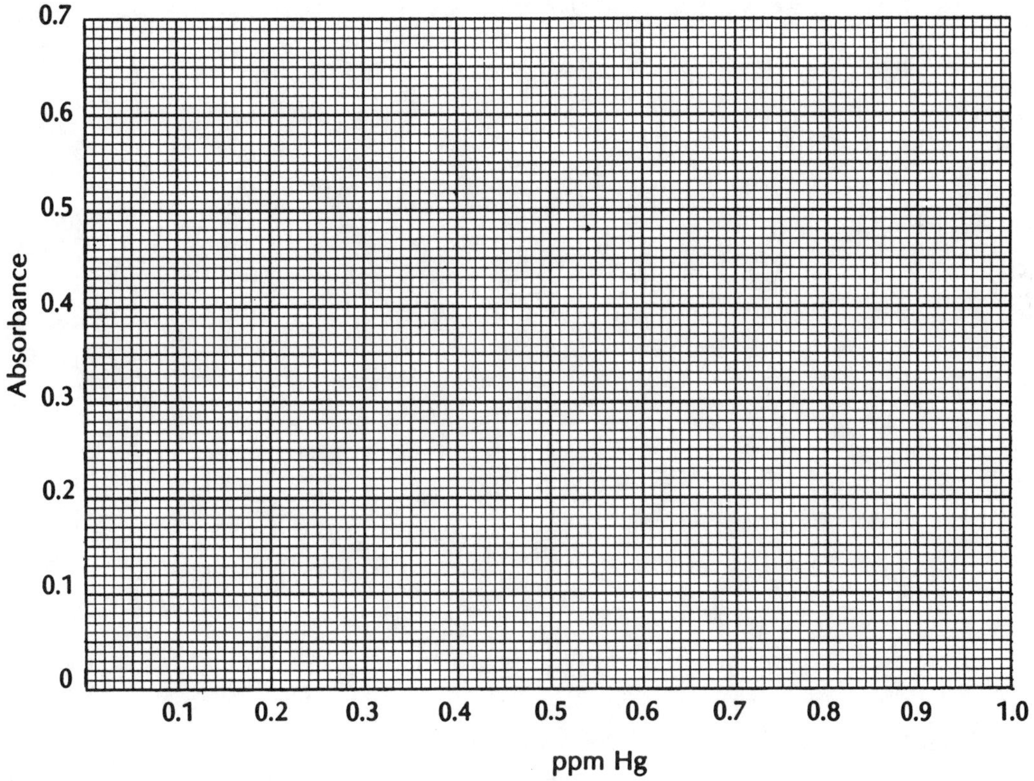

What can be said about the mercury content of the tap water?

Experiment 21
Analysis of the Amount of Phosphate in Water

Object

The amount of phosphate in samples of water will be determined by spectrophotometry.

Background

While phosphates occur in nature in many forms, relatively little of these phosphates find their way into water supplies. Soaps and detergents, industrial wastes, and fertilizers, however, have been responsible for the introduction of large amounts of phosphates into lakes, streams, ponds and other water supplies often with disastrous ecological results. Phosphates act as a fertilizer for the plant life in the water. Furnished with this extra fertilizer, the plants grow so rapidly that they fill the water supply and figuratively choke it to death. This process is called eutrophication. A high phosphate in drinking water gives the water a soapy taste.

In this experiment you will determine the amount of phosphate in two water samples, one of which is a prepared unknown and one is a sample collected by you from a lake or other drainage water source.

To analyze the sample, a spectrophotometric (colorimetric) method is used. A reagent is added that reacts specifically with phosphate to produce a colored solution. The amount of phosphate present is directly proportional to the intensity of color. The procedure involves making up a series of phosphate solutions of known concentration and then adding the coloring agent. The intensity of color is then measured for each solution. This could be done visually, but is more accurately done using an instrument called a spectrophotometer (or colorimeter). A calibration plot is made of phosphate concentration versus intensity of color (absorbance). The more phosphate that there is in the solution, the more intense will be the color. The unknown samples are then treated with the coloring agent and their color intensity measured. By using the calibration plot, the concentration of phosphate is determined. One simply finds what phosphate concentration in the unknown sample corresponds to that absorbance.

Experimental Procedure

1. The theory and use of the spectrophotometer is discussed in a section at the front of this manual.

   a. Obtain your colorimeter and colorimeter test tubes.

   b. Turn on the instrument and adjust the wavelength-control knob to 660 nanometers. Allow the instrument to warm up for 10 minutes.

   c. With the sample compartment empty, adjust the zero-control knob (the left-hand knob) so that the meter pointer reads infinite absorbance. (This is the same as zero percent transmittance.) Close the cover of the sample holder when you are adjusting the meter.

   d. Fill one colorimeter test tube with distilled water and place it in the sample compartment of the instrument. Close the cover. The meter pointer should move to the right.

   e. Using the percent-transmittance knob (the right-hand knob), adjust the meter pointer to read zero absorbance. (This is the same as 100 percent transmittance.)

f. The instrument is now calibrated for use. You may wish to repeat the above procedure to be sure that the instrument is stable.

2. Obtain from your instructor the standard phosphate solutions of known concentration. Also obtain your unknown phosphate solution and the lake water sample.

3. Prepare these solutions in the following manner:

   a. Using your 25-ml graduated cylinder, carefully measure 25 ml of each known solution. Do the same for the unknown solutions. Pour each solution into its own 100-ml beaker.

   b. Using your small graduated cylinder, measure 5 ml of color reagent for each solution. Add this color reagent to each solution in its beaker.

   c. Stir each solution and allow the color to develop for 15 minutes.

4. Place each solution in its own colorimeter test tube and measure its absorbance in the colorimeter. Close the cover of the sample holder when you are taking a reading.

5. Construct a graph. Plot absorbance versus concentration for each of the known phosphate samples. From this graph and the absorbance of your unknown sample, determine the concentration of phosphate in your unknown samples.

Name _____   Lab Section _____

Station _____   Date _____

Analysis of the amount of Phosphate in Water

Sample	Absorbance	Phosphate Concentration (in ppm)
Known	_____	_____
Known	_____	_____
Known	_____	_____
Known	_____	_____
Known	_____	_____
Unknown	_____ (Unknown no._____)	_____
Lake	_____	_____

What could you do to obtain the concentration of phosphate in your unknown sample if the following situation occurred: You measure out 25 ml of your unknown solution. You add to your sample 5ml of color reagent. The color develops. After 15 minutes you try to read its absorbance on the colorimeter. However, you find that the color of the solution is so intense that it registers off the meter at infinite absorbance.

What was the concentration of phosphate in the lake water? What are the most likely sources of this phosphate?

133

Absorbance

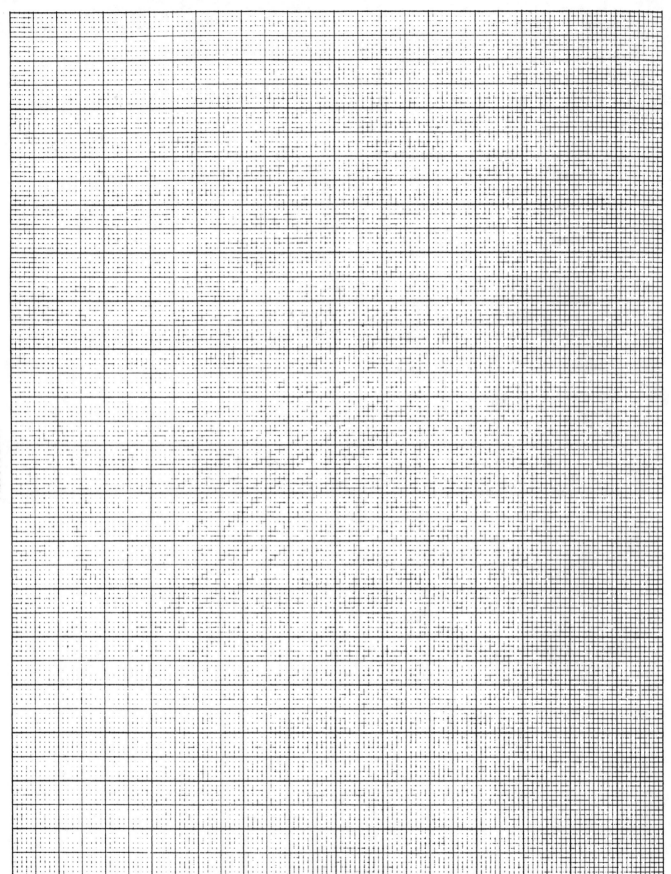

ppm Phosphate

Experiment 22
Titrations With Permanganate

Object

In this experiment you will perform several titrations using the oxidizing agent potassium permanganate in the buret. The solution will be standardizided against sodium oxalate and will then be used to measure the iron content in an unknown salt.

Background

The gram equivalent weight (GEW) of acids or bases is defined as the sample which produces one gram-mole of H^+ or OH^-. For an oxidizing or reducing agent, the GEW is defined as the sample which produces or consumes one mole of electrons. In each reaction in today's experiment, potassium permanganate, $KMnO_4$, undergoes the following reaction:

$$MnO_4^- + 8H^+ + 5e \longrightarrow Mn^{2+} + 4H_2O$$

Therefore, $KMnO_4$ is an oxidizing agent which consumes 5 moles of electrons per mole of $KMnO_4$. One equivalent of $KMnO_4$ is only one-fifth of a mole.

The $KMnO_4$ solution is standardized against sodium oxalate, $Na_2C_2O_4$. The reaction of the is:

$$C_2O_4^{2-} \longrightarrow 2CO_2 \uparrow + 2e$$

The equation shows that one mole of $Na_2C_2O_4$ contains two equivalents. A weighed $Na_2C_2O_4$ sample is dissolved to give exactly 100 ml of solution. The normality of the oxalate solution can be calculated since, normality is defined as:

$$N = \text{number of equivalents/number of liters}$$

When the oxalate solution is titrated to the end point, the number of equivalents of oxalate equals the number of equivalents of permanganate. Since normality times volume is number of equivalents, we can write:

$$N_O V_O = N_M V_M$$

Where N_O and V_O are the normality and the volume of oxalate solution used in the titration, and N_M and V_M represent the same for the permanganate solution used in the titration.

Once the permanganate is standardized (that is, its normality is determined in a direct experimental mesurement) it is used for measuring the iron content in an unknown salt. The iron is present in the form of ferrous, Fe^{2+} or $Fe(II)$, and is oxidized by permanganate by the reaction:

$$Fe^{2+} \longrightarrow Fe^{3+} + e$$

Ferrous is the only reducing agent present in the unknown salt. All permanganate in the titration of the unknown salt is consumed by reacting with ferrous in the salt, so that we can write:

$$N_M V_M = \text{number of equivalents of ferrous}$$

From the number of equivalents of iron in a salt sample, you can calculate the grams of iron and the per cent iron in the unknown salt.

A key feature of any titration is to stop the addition of one reagent exactly at the end point. At the end point, the two reagents (acid and base, or oxidizing agent and reducing agent) are chemically equivalent. In order to detect the end point in acid-base titrations an indicator is added. The indicator changes color at the end point. No indicator is necessary in a permanganate titration. Since permanganate is a deep purple color, a slight excess of permanganate can easily be detected. The end point of a permanganate titration is the point where the solution retains a faint purple color. At this point, permanganate has consumed all reducing agent in the solution being titrated.

Procedure

1. Weigh a clean, dry 50 ml beaker to ± .001 g. Add about 1 g of sodium oxalate, $Na_2C_2O_4$, and reweigh to ±.001 g. Dissolve the oxalate in distilled water and quantitatively transfer it to a clean 100 ml volumetric flask. Be sure to rinse the beaker with several small portions of distilled water to transfer all oxalate into the flask. Dilute the solution to a total volume of exactly 100 ml and mix it well. Calculate the normality of this oxalate solution.

2. Prepare 400 ml of a 0.1 N $KMnO_4$ solution. A stock solution which is 0.1 M $KMnO_4$ will be provided. Dilute the 0.1 M solution, pour it into a 500 ml erlenmeyer flask, and stir well. Clean and dry a 50 ml buret. Rinse it with several portions of the permanganate solution. Discard the permanganate used for rinsing. Fill the buret with permanganate solution and drain the buret back into the 500 ml flask several times to make sure all of the permanganate has a uniform concentration.

3. Pipete 20 or 25 ml of oxalate into a clean 250 ml erlenmeyer flask. Use a dry pipete if possible. If a dry pipete is not available, dry the outside of a pipete and rinse the inside with several small portions of the oxalate solution first. Remember to use the rubber bulb to fill the pipete. Oxalate is quite toxic — **DO NOT** pipete by mouth. Add 20 ml of 3 M H_2SO_4 and 50 ml of distilled water to the erlenmeyer flask.

4. Fill the buret; read the initial permanganate solution level to ± .01 ml using the top of the solution. Permanganate solutions may be so dark that the meniscus is not visible. Add about 20 ml of permanganate to the oxalate solution. Heat the solution to about 80°C. Continue the addition of permanganate to the end point. Add permanganate slowly near the end point. Dropwise addition is suggested at the last. The end point is a faint purple or pink color which will last 30 seconds or more. Read the buret again to ± .01 ml. Calculate the normality of the permanganate solution.

5. Pipete another oxalate sample into another erlenmeyer flask and titrate as before. Calculate the normality of permanganate from this titration. If the two normality values differ by more than 1%, perform a third titration. Calculate the average normality for permanganate by using all of your valid values.

6. Weigh a clean, dry 50 ml beaker to $\pm$.001 g. Add 1.1 to 1.3 g of unknown iron salt. Reweigh to ±.001 g. Add 20 ml of distilled water to the beaker. Quantitatively transfer all of the *salt* to a clean 250 ml erlenmeyer flask. Rinse the beaker several times to be sure all the salt has been removed. Add 10 ml of 3 M H_2SO_4 and 3 ml of 85% H_3PO_4. Fill the buret; read the permanganate level. Titrate the ferrous salt until a faint pink color will remain for 30 seconds or more. Read the permanganate level in the buret. Calculate the number of equivalents of iron in the unknown sample. You are correct in assuming that all iron in the unknown salt is present as ferrous.

7. Weigh another 1.1 to 1.3 g sample of unknown iron salt. Titrate with permanganate as before. Calculate the per cent iron in the salt. If your two values do not agree well, repeat the determination a third time.

8. Average the valid values of %Fe in the unknown.

Name_____   Section_____

Station_____   Date_____

Titrations With Permanganate

Standarization of Permanganate Solution

Weight of beaker,_____ g; weight of beaker + $Na_2C_2O_4$,_____ g

Weight of $Na_2C_2O_4$,_____ g; total volume of oxalate solution_____ml.

Normality of oxalate solution,_____N. Show your work.

Preparation of $KMnO_4$ solution:_____ml of_____ M $KMnO_4$ diluted to_____ml

total volume. $KMnO_4$ solution is about_____ M or_____ N.

Trial 1:_____ml of $Na_2C_2O_4$ solution; initial buret reading,_____ml; final buret reading,

_____ml. Volume of permanganate,_____ml. Normality of Permanganate, this

trial,_____N. Show your work.

Trial 2:

Trial 3:

Trial 4:

Average normality of $KMnO_4$ solution is_____N. Show your work.

Name_____   Section_____

Station_____   Date_____

Titrations With Permanganate

Standarization of Permanganate Solution

Weight of beaker,_____ g; weight of beaker + $Na_2C_2O_4$,_____ g

Weight of $Na_2C_2O_4$, _____ g; total volume of oxalate solution_____ml.

Normality of oxalate solution,_____N. Show your work.

Preparation of $KMnO_4$ solution: _____ml of_____ M $KMnO_4$ diluted to_____ml

total volume. $KMnO_4$ solution is about_____ M or_____ N.

Trial 1: _____ml of $Na_2C_2O_4$ solution; initial buret reading, _____ml; final buret reading,

_____ml. Volume of permanganate,_____ml. Normality of Permanganate, this

trial,_____N. Show your work.

Trial 2:

Trial 3:

Trial 4:

Average normality of $KMnO_4$ solution is_____N. Show your work.

Iron Content in Salt

Trial 1: Weight of beaker, _____ g; weight of beaker plus unknown salt, _____g.

Weight of salt, _____g. Initial buret reading, _____ml;

Final buret reading, _____ml; volume of $KMnO_4$, _____ml.

Equivalents of iron in sample, _____.

Grams of iron in sample, _____.

Per cent iron in sample, _____. Show your work.

Trial 2:

Trial 3:

Average value of %Fe in sample, _____. Show your work.

Experiment 23
Determination of Calcium Content in Hard Water

Object

The amount of calcium in the local water supply, which is responsible for hardness, will be determined by titration with the chelating agent ethylenediaminetetraacetic acid, EDTA.

Background

Hard water is water that contains dissolved salts of the metal ions calcium, magnesium and iron. These ions react with soap to give a precipitate. In this area our hardness is due primarily to calcium which enters our water supply when ground water percolates through limestone (caliche, $CaCO_3$). Hard water causes many problems beyond the formation of "bathtub ring", primarily relating to scale formation.

The concentration of many metal ions can be determined in solution by the use of complexometric titration. In this experiment an indicator (Eriochrome Black T) is added to a sample of tap water. The indicator forms a *red* complex with the metal ions.

$$\text{excess } M^{2+} \ + \ \text{indicator} \longrightarrow [M\text{-indicator}]^{2+} \ + \ M^{2+}$$

$$\text{metal ion} \qquad\qquad \text{blue} \qquad\qquad\qquad \text{red}$$

Many more metal ions are present in the water than will complex with the small amount of indicator that is added. Thus most of the metal ions remain as uncomplexed or "free". The metal ions form a more stable complex with the complexing agent EDTA

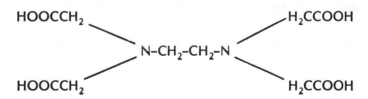

As EDTA is added (titrated) to the water-indicator solution, the EDTA removes or complexes with the free metal ions.

$$\text{free } M^{2+} \ + \ \text{EDTA} \ + \ [M\text{-indicator}]^{2+} \longrightarrow M\text{-EDTA}^{2+} \ + \ [M\text{-indicator}]^{2+}$$

As the endpoint in the titration is reached, all of the free M^{2+} ions have been removed (complexed in the form of $M\text{-EDTA}^{2+}$), leaving only M-indicator^{2+}. The EDTA then displaces the indicator from the M-indicator^2 ions resulting in free indicator and a color change.

$$[M\text{-indicator}]^{2+} \ + \ \text{EDTA} \longrightarrow M\text{-EDTA}^{2+} \ + \ \text{free indicator}$$

$$\text{red} \qquad\qquad\qquad\qquad\qquad\qquad\qquad \text{blue}$$

The essential feature in this reaction is that one EDTA molecule is used for each M^{2+} metal ion in the original water. By knowing the concentration of the EDTA solution that is used in the titration, the concentration of metal ions in the water can be calculated.

The calculation will determine the total amount of hard metal ions (Ca^{2+}, Mg^{2+}, Fe^{3+}) in the water. The local water supply mainly contains Ca^{2+}, however.

(Molarity of EDTA) × (Volume of EDTA) = (Molarity of ions) × (Volume of water sample).

The only unknown in the equation is the molarity of the ions in the water sample. Solving the equation will tell you the number of moles of metal ions per liter of water, that is, the molarity. If you assume all of the hardness is due to calcium ions in the form of calcium carbonate, $CaCO_3$, then the concentration of $CaCO_3$ can also be calculated in terms of grams per liter of $CaCO_3$ using the relationship

grams = (number of moles) × (Molecular weight).

The units ppm is equal to milligrams per liter.

Experimental Procedure

1. Using a 50-ml pipette, measure 50 ml of *tap water* into a 250-ml Erlenmeyer flask. Add 10 ml of buffer solution (to bring the pH to a value of about 10). Swirl the solution gently while adding sufficient indicator (Eriochrome Black T) to get an easily visible color (10 to 15 drops).

2. Rinse a clean 50-ml buret with several small amounts of the standard Na_2EDTA solution. Fill the buret to the zero ml mark.

3. Titrate the tap water-buffer-indicator solution with the EDTA solution until the solution just changes from red to purple to blue. This should be done by adding small amounts of the Na_2EDTA solution to the tap water solution, swirling the mixture until the solution changes to a uniform color. Near the endpoint, the Na_2EDTA should be added dropwise. When the color of the solution is a uniform blue, stop the titration and record the volume of Na_2EDTA standard solution used.

4. Repeat the titration on a second sample prepared as in step 1, filling the buret back up to the zero ml mark.

Name_____ Lab Section_____ _____

Station_____ Date_____ _____

Determination of Calcium Content in Hard Water

Na_2EDTA solution   Titration 1   Titration 2

(1) Initial reading   _____ ml   _____ ml

(2) Final reading   _____ ml   _____ ml

(3) Volume used   _____ ml   _____ ml

(4) Average volume used   _____ ml

Concentration of standard Na_2EDTA used   _____ molar

Volume of tap water used   _____ ml

Molarity of metal ions in tap water — show your calculations:

_____ molar

Concentration of metal ions in tap water in units of grams per liter (assuming all metal ions are due to $CaCO_3$). Show your work:

_____ g/l

Concentration of metal ions in tap water in units of ppm. Show your work:

_____ ppm

145

What species is responsible for the blue color?

For the red color?

Experiment 24
Heat of Reaction

Object

In today's experiment you will construct a simple calorimeter and use it to measure the heat associated with several chemical reactions. The calorimeter will be constructed from styrofoam cups, a thermometer, a lid and a wire stirrer.

Background

For today's experiment you will construct a simple but effective calorimeter from several parts. Nested styrofoam cups and a lid will be used to minimize heat loss from the calorimeter. A thermometer will be used to measure temperatures to $\pm 0.1°$. The contents will be stirred to insure a uniform temperature inside the calorimeter. This calorimeter will be used to measure the heat produced from several acid-base reactions. The measurements will be used to calculate the calories of heat produced in the experiment. The heat of reaction (the enthalpy change) for the reaction will be obtained by calculating the kilocalories of heat per mole of reaction.

In all parts of the experiment, the calorimeter will contain 100 mL of water (or 100 mL of an aqueous solution). Heat produced by the reaction will warm the 100 mL of water plus the calorimeter parts. The specific heat of water is 1.00 cal/(g-deg); therefore it requires 100 cal to warm the water by one degree. If we designate the heat capacity of the calorimeter parts as C, it follows that it will require $(100 + C)$ calories to warm up the water and the calorimeter parts by one degree. In any experiment, the calories of heat produced will be equal to $(100 + C)$ times the temperature change.

In Part 1 you will mix 50 ml of hot water with 50 mL of water at room temperature. The results will be used to obtain a value of C. In Part 2 you will mix 50 mL of 1.00 M HCl with 50 mL of base. In Part 3 you will mix 50 mL of 2.00 M HCl with 50 mL of base. In Part 4 you will mix 50 mL of 2.00 M acetic acid, $HC_2H_3O_2$, with 50 mL base. In each case the acid will be the limiting reagent. You will measure the calories of heat produced by each mixture. You will calculate the heat of reaction or enthalpy change produced by each reaction. You will then deduce the heat of reaction for the ionization of acetic acid from your experimental data.

The number of calories required to raise the internal parts of the calorimeter by $1°$ Celsius is designated as C. The value of C will be measured in Part 1; in this part you will initially place 50 mL of water in the calorimeter. You will measure the initial temperature of the water and calorimeter parts (designated T_1) at the beginning of the experiment. You will heat a second 50 mL water sample to a temperature measured as T_2. The 50 mL sample of warm water is poured into the calorimeter, and the final temperature of the 100 mL of water and calorimeter parts is measured as T_f. You can assume that the heat lost from the hot water is equal to the heat gained by the cool water and the calorimeter parts. The heat lost by the warm water is calculated as the heat capacity, 50 cal/deg, times the temperature change, $T_2 - T_f$ in equation form it is written:

$$\text{Heat lost} = 50 \text{ cal/deg} \times (T_2 - T_f) \text{ deg}$$

The heat gained by both the calorimeter parts and the 50 mL of water is calculated as the heat capacity, $50 + C$, times the temperature change, $T_f - T_1$. In equation form it is written:

$$\text{Heat gained} = (50 + C) \text{ cal/deg} \times (T_f - T_1) \text{ deg}$$

When T_1, T_2, and T_f are measured experimentally, we can calculate heat lost, set it equal to heat gained, and then solve for C. In the remaining measurements you can assume that the heat produced by the chemical reaction is equal to $(100 + C)$ times the temperature change.

In Part 2 of the experiment, you will measure the heat associated with the reaction:

$$NaOH(aq) + HCl(aq) \longrightarrow NaCl(aq) + H_2O(1)$$

You will place 50 mL of 2.2 M NaOH (which contains 0.11 moles of NaOH) in the calorimeter. You will then prepare 50 mL of 1.0 M HCl solution, and make sure that both solutions are at the same temperature, T_3. The acid is then poured into the calorimeter and the liquid is mixed. The reaction will take place and will cause the temperature to rise to T_4. The heat produced from the reaction, Q, is calculated from the equation:

$$Q = (100 + C) \text{ cal/deg} \times (T_4 - T_3) \text{ deg}$$

Notice that in Part 2 you mixed 0.11 moles of NaOH and 0.050 moles of HCl. From the balanced chemical equation it is clear that the reaction will produce 0.050 moles of NaCl and then all the HCl is consumed. HCl is the limiting reagent. Results of calorimetry experiments should be converted into heat of reaction values which express the heat per mole of reaction. For this calculation the Q should be converted into kcal and then divided by the moles of limiting reagent.

$$\text{Heat of reaction} = \frac{Q \text{ cal } (1 \text{ kcal}/1000 \text{ cal})}{\text{moles HCl}}$$

In Part 3 of the experiment you will place 50 mL of 2.2 M NaOH in the calorimeter again and add 50 mL of 0.20 M HCl. The calculations are similar to those of Part 2, except that 0.10 mole of the limiting reagent, HCl, is used in this experiment.

Remember that HCl is a strong acid and NaOH is a strong base. This means that in solution, HCl is completely dissociated into H^+ and Cl^- ions and NaOH is completely dissociated into Na^+ and OH^- ions. The Cl^- and Na^+ ions do not react, so we can write the net reaction as:

$$H^+ + OH^- \longrightarrow H_2O$$

Any strong acid should give the same value for the heat of reaction. Acetic acid is a weak acid which is only partially ionized in solution. When it reacts with base we can write the reaction as:

$$HC_2H_3O_2 + OH^- \longrightarrow H_2O + C_2H_3O_2^-$$

You will measure this heat of reaction in Part 4. You can consider it to be the sum of the two reactions:

$$HC_2H_3O_2 \longrightarrow H^+ + C_2H_3O_2^-$$

$$H^+ + OH^- \longrightarrow H_2O$$

148

By using the rule that heats of reaction can be added just like chemical equations, you will use your experimental values to determine the heat of reaction for the ionization of acetic acid.

Procedure

Perform this experiment with a lab partner.

1. Assemble the calorimeter as shown in figure 24-I. Measure 50 mL of distilled water in a graduated cylinder and pour this into the calorimeter. Measure another 50 mL of water and place this in a beaker. Heat this sample to about 50-55° C. Borrow a second thermometer for a few minutes. Place both thermometers into the beaker of warm water. Stir well and determine the reading on the borrowed thermometer which corresponds to 50.0° on your thermometer. Move your thermometer back to the calorimeter. One lab partner should stir the calorimeter for at least five minutes; note this temperature as T_1. The other lab partner should heat and stir the beaker to obtain the required temperature on the borrowed thermometer. When the beaker contents are at the temperature which corresponds to 50.0° on your thermometer (but the actual measurement is made with the borrowed thermometer), pour the hot water into the calorimeter. This should make $T_2 = 50.0°$. Stir the calorimeter contents well and then read the temperature at fifteen second intervals. The highest temperature should occur within two minutes. Record this temperature as T_f. Calculate the heat lost by the hot water and the value of C, the heat capacity of the calorimeter parts.

2. Dry the calorimeter and reassemble it. Measure out 50 mL of 2.2 M NaOH and pour it into the calorimeter. Use a clean dry graduated cylinder to measure 25 mL of 2.0 M HCl. Fill the cylinder to the 50 mL mark with distilled water. Mix this solution well. Measure the temperature of both solutions, to be sure that they are at the same temperature. Record this temperature as T_3. Pour the HCl into the calorimeter; stir the solution and read temperatures at fifteen second intervals. Read the highest temperature as T_4. Test the calorimeter solution with phenolphthalein. Calculate the calories of heat produced by the reaction. Calculate the heat of reaction as kcal of heat per mole of HCl.

3. Follow the same directions as in Part 2, except that the acid solution should contain 50 mL of 2.0 M HCl. Be sure to test the final solution with phenolphthalein.

4. Follow the same directions as in Part 2, except that the acid solution should contain 50 mL of 2.0 M $HC_2H_3O_2$. Calculate the calories of heat produced by the reaction and the heat of reaction in kcal per mole of acetic acid. Use your data to calculate the heat of reaction for the process:

$$HC_2H_3O_2 \longrightarrow H^+ + C_2H_3O_2^-$$

you may assume that the reaction which occurred in the calorimeter is:

$$HC_2H_3O_2 + OH^- \longrightarrow H_2 + C_2H_3O_2^-$$

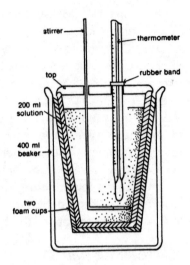

Figure 24-I. Calorimeter

Name_____   Lab Section _____

Station_____   Date_____

Heat of a Reaction

1. Temperature of cool water and calorimeter, T_1 _____° C.

   Temperature of warm water, T_2 _____° C.

   Final temperature of solution, T_f _____° C.

   Heat lost by warm water is _____ cal. Show your work.

   Heat capacity of calorimeter parts is _____ cal/deg. Show your work.

   To heat calorimeter and 100 ml of solution requires _____ cal/deg.

2. 50 ml of _____ M NaOH plus 50 ml of _____ M HCl.

   Temperature of NaOH and HCl solutions before reaction, _____° C.

   Temperature of solution after reaction, _____° C.

   Phenolphthalein is _____ in the calorimeter solution.

   Heat released in the reaction is _____ cal. Show your work below.

   Heat of reaction, expressed as kcal/mole of HCl is _____ kcal/mole.
   Show your work below.

3. 50 ml of _____ M NaOH plus 50 ml of _____ M HCl.

   Temperature of NaOH and HCl solutions before reaction, _____° C.

   Temperature of solution after reaction, _____° C.

   Phenolphthalein is _____ in the calorimeter solution.

   Heat released in the reaction is _____ cal: expressed as kcal/mole of HC1 the value is

   _____ kcal/mole. Show your work below.

151

4. 50 ml of _____ M NaOH plus 50 ml of _____ M $HC_2H_3O_2$.

Temperature of NaOH and $HC_2H_3O_2$ solutions is, _____° C.

Temperature of solution after reaction, _____° C.

Phenolphtalein is _____ in the calorimeter solution.

Heat released in the reaction is _____ cal: expressed as kcal/mole of $HC_2H_3O_2$ the

value is _____ kcal/mole. Show your work below.

Best value for the reaction: $H^+ + OH^- \longrightarrow H_2O$ is _____ kcal/mole.

Value for the reaction: $HC_2H_3O_2 + OH^- \longrightarrow H_2O + C_2H_3O_3^-$ is _____ kcal/mole.

Value for the reaction $HC_2H_3O_2 \longrightarrow H^+ + C_2H_3O_2^-$ is _____ kcal/mole. This

reaction is endothermic/exothermic. Show your work below.

152

Experiment 25
Measuring the Order of a Reaction

Object

In today's experiment you will measure the order in iodide ion and in hydrogen peroxide for the iodide ion catalyzed decomposition of hydrogen peroxide.

Background

Hydrogen and oxygen form two substances — water and hydrogen peroxide. Water is very stable, but hydrogen peroxide slowly decomposes into water and oxygen. The reaction can be written $2H_2O_2(\ell) \longrightarrow 2H_2O(\ell) + O_2(g)$ to show that a gas, O_2, is given off by this reaction. This reaction is very slow in clean glass containers, but many substances can cause the reaction to occur rapidly. Iodide ion is one catalyst for this reaction.

The rate of the above reaction can be defined in terms of the volume of O_2 generated per minute. The rate law for the reaction as catalyzed by iodide ion can be written in the form

$$\text{Rate } (^{ml}O_2/\text{minute}) = k[H_2O_2]^a[I^-]^b$$

The exponent a is the order in peroxide; the exponent b is the order in iodide. If a is one, the reaction is first order in peroxide; doubling the concentration of peroxide will double the rate of the reaction. If a is two, doubling the concentration of peroxide will increase the rate of the reaction by a factor of four. The order in a substance cannot be obtained from the chemical equation. Iodide will affect the speed of the reaction, but iodide does not appear in the chemical equation. The order in a substance must be determined by experiments.

You will utilize the equipment in Figure 25-I to measure the rate of reaction for several solutions. By studying the change in rates of reaction produced by different concentrations of H_2O_2 and I^-, you will determine the order in each reagent. The apparatus is set up so that each ml of O_2 gas produced by the reaction will cause a ml of water (liquid) to be pushed out of the apparatus. You will graph the volume of water collected against time and use this to obtain the rate of reaction. The rate may slow down as the peroxide is decomposed. You should use your data to obtain the initial rate of the reaction.

Hydrogen peroxide and water are placed in the flask. The apparatus is sealed and then the KI solution is added. The clamp is removed and the flask is swirled to mix the contents. Volume vs. time is recorded until 16 ml of water have been collected. The flask is emptied, rinsed, and refilled with a solution of different composition. The solutions to be studied are tabulated below.

Solution	3% H_2O_2	water	1 M KI
A	5 ml	14 ml	1 ml
B	10 ml	9 ml	1 ml
C	5 ml	13 ml	2 ml

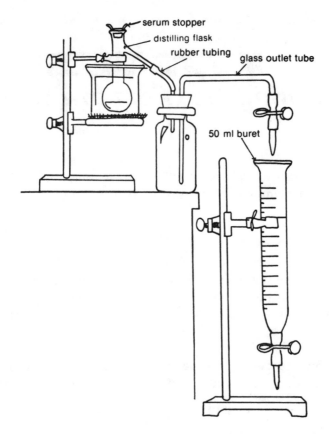

Figure 25-I. Apparatus for measuring order of a reaction

Procedure

Students should perform this experiment in pairs.

1. Set up the apparatus diagramed in Figure 25-I. Be careful when pushing glass tubing into rubber stoppers. The rubber tubing connecting the 50 ml distilling flask and the 250 ml bottle (a 250 ml flask may be used here) should be long enough to permit the flask to be shaken. Make sure the outlet tube is filled with water before you attach the pinch clamp.

2. Pipette 5 ml of 3% H_2O_2 (about 0.9 M) into the flask. Use a transfer pipette to add 14 ml of water to the flask. Attach the septum stopper. Make sure the buret has water up to the 50 ml mark. Inject 1 ml of 1 M KI solution into the flask. Remove the pinch clamp. Swirl the contents of the flask. Record the exact time (±1 second or better) when the buret has water at the 48.00 ml mark. Record times at which successive 2 ml portions of water are collected, until 16 ml of water have been collected. In case you use an ordinary watch for timing, record the actual clock reading as raw data. Convert to elapsed time assuming zero time corresponds to the first reading.

3. Empty the flask and rinse it. Pipette 10 ml of 3% H_2O_2 and 9 ml of water into the flask. Proceed using the directions in step 2 until you have collected 16 ml of water.

4. Empty the flask, and rinse it. Pipette 5 ml of 3% H_2O_2 and 13 ml water into the flask. Repeat the directions in Step 2 except that you should inject 2 ml of 1 M KI into the flask.

5. Prepare a graph of the results of each experiment. Graph volume vs. time for each solution. Connect the data points with a smooth line. Estimate the rate of reaction at zero time for each solution. The rate of reaction (ml/sec) is the slope of the line at zero time.

6. Tabulate the initial reaction rates and the solution concentrations. Use these data to determine the order of the reaction in peroxide and the order of the reaction in iodide.

Name _____   Section _____

Station _____   Date _____

Measuring the Order of a Reaction

Solution A: _____ ml H_2O_2, _____ ml H_2O plus _____ ml KI

Buret Reading ____ ____ ____ ____ ____ ____ ____ ____ ____

Clock Reading ____ ____ ____ ____ ____ ____ ____ ____ ____

Elapsed Time ____ ____ ____ ____ ____ ____ ____ ____ ____

Volume collected ____ ____ ____ ____ ____ ____ ____ ____ ____

Solution B: _____ ml H_2O_2, _____ ml H_2O plus _____ ml KI

Buret Reading ____ ____ ____ ____ ____ ____ ____ ____ ____

Clock Reading ____ ____ ____ ____ ____ ____ ____ ____ ____

Elapsed Time ____ ____ ____ ____ ____ ____ ____ ____ ____

Volume Collected ____ ____ ____ ____ ____ ____ ____ ____ ____

Solution C: _____ ml H_2O_2, _____ ml H_2O plus _____ ml KI

Buret Reading ____ ____ ____ ____ ____ ____ ____ ____ ____

Clock Reading ____ ____ ____ ____ ____ ____ ____ ____ ____

Elapsed Time ____ ____ ____ ____ ____ ____ ____ ____ ____

Volume Collected ____ ____ ____ ____ ____ ____ ____ ____ ____

Initial rate of reaction in Solution A, _____ ml/sec

Show your work below.

Initial rate of reaction in Solution B, _____ ml/sec

Initial rate of reaction in Solution C, _____ ml/sec

Summary of Data

Solution	ml H_2O_2	ml KI	Total Volume of Soln	Initial Rate
A	_____	_____	_____	_____
B	_____	_____	_____	_____
C	_____	_____	_____	_____

The reaction is _____ order in H_2O_2. State your reasons below.

The reaction is _____ order in iodide. State your reasons below.

Experiment 26
Molecular Models

Object

Molecules are three dimensional: however, it is important to be able to represent three dimensional structures in two dimensions, i.e., on paper. The purpose of today's experiment is to develop the facility for visualizing three dimensional molecules from two dimensional formulae.

Background

The ability of carbon to form very strong bonds with itself leads to the possibility of a large number of isomers differing only in the spatial arrangements of the atoms comprising the molecules. One must be able to recognize the three dimensional structure of a molecule from a two dimensional projection drawing. For example, dichloromethane can be drawn as

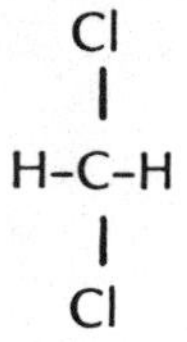

$$\begin{matrix} & Cl \\ & | \\ H- & C & -H \\ & | \\ & Cl \end{matrix} \quad \text{or as} \quad \begin{matrix} & H \\ & | \\ H- & C & -Cl \\ & | \\ & Cl \end{matrix}$$

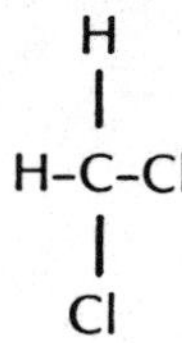

: both drawings are correct and represent only one molecule. However, 1,2-dichloroethene exists as two isomers, the *cis* isomer

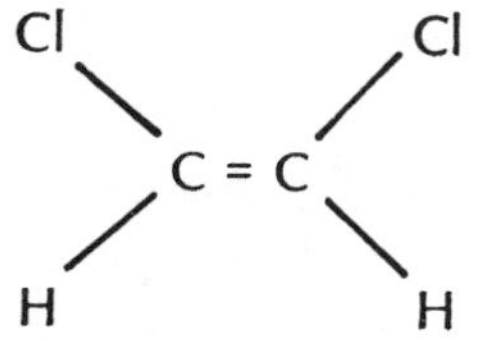

$$\begin{matrix} Cl & & Cl \\ \searrow & & \swarrow \\ & C=C & \\ \nearrow & & \nwarrow \\ H & & H \end{matrix}$$

and the *trans* isomer

$$\begin{matrix} H & & Cl \\ \searrow & & \swarrow \\ & C=C & \\ \nearrow & & \nwarrow \\ Cl & & H \end{matrix}$$

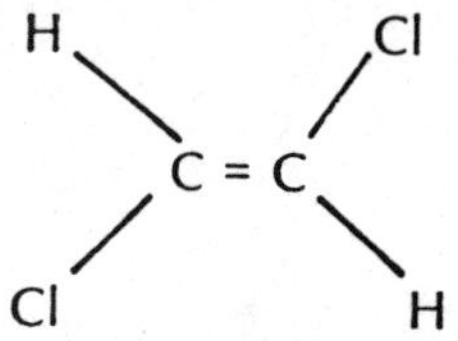

In this case the drawings represent two different molecules.

Procedure

Use the molecular model kit provided for the construction of models of all compounds having the formulae listed. After the models have been constructed, draw projection formulae for each compound. Using conventional nomenclature, name each compound.

159

Name_____ Lab Section_____ - _____

Station_____ Date_____ . ____

Molecular Models

Projection Formula   Name

I.   The Alkanes C_nH_{2n+2}

A. CH_4

B. CH_3Cl

C. CH_2Cl_2

D. CH_3CH_3

E. $CH_3CH_2CH_3$

F.   Dichloropropanes having the
     formula $C_3H_6Cl_2$ (4 isomers)

G.   Compounds having the
     formula C_5H_{12} (3 compounds)

161

Projection Formula   Name

II. The Cycloalkanes C_nH_{2n}

    A. Cyclohexane

    B. The 1,1- and 1,4-
        dichlorocyclohexanes

III. The Alkenes and alkynes C_nH_{2n}
     and C_nH_{2n-2}

    A. 1-Butene C_4H_8

    B. The 2-butenes C_4H_8
        (two isomers)

    C. The dichloroethenes $C_2H_2Cl_2$
        (three isomers)

    D. 1,3-butadiene, C_4H_6

    E. 1-butyne and 2-butyne C_4H_6

Name_____   Lab Section_____

Molecular Models

<u>Projection Formula</u>   <u>Name</u>

IV. The Aromatics

    A. Benzene C_6H_6

    B. The dichlorobenzenes
        $C_6H_4Cl_2$ (3 isomers)

V. Compounds Representing Important Functional Groups

    A. Alcohols R-OH

        1. Methanol (Wood Alcohol) CH_3OH

        2. Ethanol (Grain Alcohol) C_2H_5OH

    B. Aldehydes $R-\overset{\overset{\displaystyle O}{\|}}{C}-H$

        1. Methanal (Formaldehyde) CH_2O

        2. Ethanal (Acetaldehyde) CH_3CHO

    C. Carboxylic acids $R-\overset{\overset{\displaystyle O}{\|}}{C}-OH$

        1. Formic acid HCOOH (The bite of the red ant)

163

Projection Formula | Name

2. Acetic acid CH_3COOH (vinegar)

D. Ketones $R-\overset{\overset{O}{\|}}{C}-R$

1. Propanone (Acetone) CH_3COCH_3

2. 2-butanone (methyl ethyl ketone- M.E.K.) $CH_3COCH_2CH_3$

E. Ether R-O-R

Ethyl ether (diethyl ether)
$CH_3CH_2OCH_2CH_3$

F. Amines RNH_2 or R_2NH or R_3N

1. Ethyl amine $CH_3CH_2NH_2$

2. Diethyl amine $(CH_3CH_2)_2NH$

G. Esters $R-\overset{\overset{O}{\|}}{C}-O-R$

CH_3COOCH_3

H. Amides $R-\overset{\overset{O}{\|}}{C}-N-R'$
Acetamide CH_3CONH_2

VI. Optical Isomers

1. CHBrC1I

Experiment 27
Valence Shell Electron Pair Repulsion Theory

Object

Valence Shell Electron Pair Repulsion (VSEPR) Theory will be used to predict the structures of a number of molecules and ions. Ballon models will be constructed to verify the prediction of structural types.

Background

VSEPR (Valence Shell Electron Pair Repulsion) Theory is an extremely simple, easily conceptualized, yet powerful theory which allows one to predict with great accuracy the structure of thousands of molecules and ions. The theory is limited to the main group elements (that is, the theory is not applicable to the transition metals). In simplist terms, the theory states that the structure of a molecule or ion is determined by the number of pairs of electrons around the *central atom.* Since electrons are all of the same charge, then the theory states that the electron *pairs* will tend to separate themselves as far as possible.

In order to use VSEPR Theory, all that is necessary is to draw a Lewis (electron dot) structure of a molecule or ion. The total number of electron pairs around the central atom is counted, and from this information, the structure of the species can be determined.

In today's experiment, you will use ballons to represent pairs of electrons. The ballons (electron pairs) when tied together will automatically assume a shape which minimizes the ballon-ballon (electron pair-electron pair) interaction. The shape of the molecule or ion is determined by the number of ballons that are tied together (that is, the number of electron pairs around the central atom).

Before using the ballons to determine structures, the use of Lewis dot structures will be reviewed. This review is an abbreviated presentation which will concentrate on results rather than theory. In order to determine a Lewis dot structure, it is generally assumed that every element will, if possible, achieve an electron configuration of the closest rare gas — an octet of electrons. (For hydrogen, the closest rare gas is helium, which has two electrons. That is, an "octet" of electrons for hydrogen is two).

To draw a Lewis (electron dot) structure:

1. Assign the central atom the same number of dots (electrons) as the Group that the element is in. The Group number, then, is the same as the number of valence electrons. Note: the unique atom, if there is a unique atom, will almost always be the central atom.

2. Around the central atom, place the other atoms, again assigning the number of dots (electrons) to these atoms as the Group that they are in the periodic chart.

3. Two electrons shared between two atoms (that is, a chemical bond) will apply towards the total electron count around both of the bonded atoms. If two atoms A and B both have one electron each that are shared in an electron pair (a bond), then the new electron count for each atom will

165

increase by one electron, that is, the shared electron. Electrons are shared until the electron count for each atom is eight (an octet, or for hydrogen, two).

4. Any electrons around the central atom that are not involved in a shared pair of electrons (a bond), are paired together and become a "lone pair", that is, an "unshared pair" of electrons. All electrons around the central atom will be paired, either bonding pairs or lone pairs. The result of the above operations is the Lewis dot structure of the molecule.

Example: draw the Lewis dot structure for CCl_4.

1. The unique atom carbon C is assumed to be the central atom. Therefore, since C is in Group IV, we draw four dots (electrons) around C.

$$\cdot \overset{\displaystyle \cdot}{C} \cdot$$

2. The four Cl atoms are placed around the C atom. Since Cl is in Group VII in the periodic chart, then we place seven dots (electrons) around each Cl.

$$:\overset{\displaystyle ..}{\underset{\displaystyle ..}{Cl}}:$$
$$:\overset{..}{\underset{..}{Cl}}:\cdot\cdot\overset{..}{C}\cdot\cdot:\overset{..}{\underset{..}{Cl}}:$$
$$:\overset{\displaystyle ..}{\underset{\displaystyle ..}{Cl}}:$$

3. If each Cl atom shares one electron with the C atom, then each Cl will gain one electron through sharing, and will therefore complete its octet; furthermore, the C atom will gain one electron for each of the Cl atoms with which it is sharing a pair of electrons. Thus the C atom will have it's octet of electrons: four that it originally had plus one each from the four Cl atoms to which it is bonded. The Lewis dot structure then appears as:

$$:\overset{..}{\underset{..}{Cl}}:$$
$$:\overset{..}{\underset{..}{Cl}}:\overset{..}{C}:\overset{..}{\underset{..}{Cl}}:$$
$$:\overset{..}{\underset{..}{Cl}}:$$
, which can also be drawn as

$$:\overset{..}{\underset{..}{Cl}}:$$
$$:\overset{..}{Cl}-\overset{..}{C}-\overset{..}{Cl}:$$
$$:\overset{..}{\underset{..}{Cl}}:$$
, or simply as

$$\overset{Cl}{\underset{Cl}{\overset{|}{Cl-\overset{|}{C}-Cl}}}$$

The dot structure for CCl_4, then, has four bonding pairs of electrons around the central C atom.

Example: draw a Lewis dot structure for ammonia NH_3.

1. Nitrogen is assumed to be the central atom. Nitrogen is in Group V, so five valence electrons are drawn around it.

$$\cdot \overset{\displaystyle ..}{\underset{\displaystyle .}{N}} \cdot$$

2. The three H atoms are drawn around N, and since H is Group I in the periodic chart, one electron (dot) is drawn by each H atom.

$$H\cdot\cdot\overset{\displaystyle ..}{\underset{\displaystyle .}{N}}\cdot\cdot H$$
$$H$$

166

3. Nitrogen is three electrons shy of it's octet, so it will share an electron with each of three H atoms, which allows N to complete its octet. Each H atom was one electron shy from it's "octet" of two electrons, so when a H atom shares an electron from nitrogen, it's rare gas configuration of two electrons is complete.

$$H:\ddot{N}:H$$
$$H$$

4. The N atom has two electrons which are not shared (bonded), so they are grouped together as a lone pair of electrons. The result then is

$$H:\ddot{N}:H \quad \text{, which is then same as} \quad H-\overset{..}{\underset{|}{N}}-H$$
$$H \qquad\qquad\qquad\qquad\qquad H$$

The N atom in NH_3, then, has three bonding pairs of electrons and one lone pair of electrons — a total of four electron pairs.

Lewis Theory for Ions. Example: draw the electron dot structure for the ammonium ion (a cation) NH_4^+.

To draw a Lewis dot structure for a cation, simply assign the charge (a positive +1, in this case) to the *central* atom by removing electrons (in this case removing one electron will result in a charge of +1). Then proceed as usual. For a negative ion (an anion), the negative charge will be assigned to the *central* atom by addition of electrons (one electron, for a charge of minus one, two electrons for a charge of minus two, etc.)

1. Nitrogen is in Group V, so five dots are placed around the central atom of N. To create the charge of +1 for the NH_4^+ cation, one electron will be removed.

$$\cdot\overset{\cdot}{\underset{\cdot}{N}}\cdot \quad \text{, then} \quad \cdot\underset{\cdot}{N}\cdot^+$$

2. The four H atoms are then placed around N with one electron each since H is in Group I.

$$\begin{array}{c} H \quad + \\ \vdots \\ \cdot H\cdot\cdot N\cdot\cdot H \\ \vdots \\ H \end{array}$$

3. Each H atom is one electron short of its "octet" of two electrons, and the N atom is four electrons short of its octet. If the nitrogen atom shares one electron each from the four H atoms, then N will achieve it's octet, and simultaneously each H atom will gain one electron satisfying it's rare gas configuration. The structure will be

$$\begin{array}{c} H \quad + \\ \vdots \\ H:N:H \\ \vdots \\ H \quad \text{, or} \end{array} \qquad \begin{array}{c} H \quad + \\ | \\ H-N-H \\ | \\ H \end{array}$$

The nitrogen atom therefore has four bonding electron pairs around it.

Lewis Theory and Multiple Bonds. Example: draw the Lewis dot structure for CO_2.

Sometimes it is necessary for bonded atoms to share more than one pair of electrons in order to complete an octet of electrons. If, as is ordinarily the case, one pair of electrons is shared between atoms, a single bond is formed; if two pairs of electrons are shared, then a double bond results as in O_2; if three pairs of electrons are shared, then a triple bond results as in N_2. (The reader should verify that O_2 has a double bond and that N_2 has a triple bond). In the case of carbon dioxide CO_2:

1. The central atom is assigned to the unique C atom, and since C is Group IV of the periodic chart, it is assigned four valence electrons.

$$\cdot \overset{\cdot}{C} \cdot$$

2. Two O atams are placed around C, and each O atom is assigned six valence electrons, since oxygen is in Group VI.

$$:\overset{\cdot\cdot}{O}:\, \cdot\overset{\cdot}{C}\cdot \,:\overset{\cdot\cdot}{O}:$$

3. The C atom is four electrons shy of it's octet, and each O atom is two electrons shy of it's octet. The only way for both oxygen atoms and the carbon atom to achieve an octet of electrons is multiple bond formation; that is, for the C and O atoms to share more than one electron pair between them. If each O shares two electrons with two electrons from C, then each oxygen gains two electrons to achieve it's octet, and the C will gain four electrons (two from each O atom). The resultant dot structure will be

$$:\overset{\cdot\cdot}{O}::C::\overset{\cdot\cdot}{O}; \text{ which is the same as } \quad O=C=O$$

In this case, then, C has two double bonds around it.

Violations of Lewis Theory: In a few molecules or ions, it is impossible to draw an octet of electrons around the central atom. Such a species is referred to as electron deficient. An example is BCl_3, where the B atom has only six electrons around it; the B atom is electron deficient. Since B is in Group III, the dot structure for BCl_3 will be

$$:\overset{\cdot\cdot}{\underset{\cdot\cdot}{Cl}}:B:\overset{\cdot\cdot}{\underset{\cdot\cdot}{Cl}}: \quad\quad \text{or} \quad\quad Cl-\underset{\underset{Cl}{|}}{B}-Cl$$

The compound boron trichloride, BCl_3, is extremely reactive, since it will react with almost any species from which it can gain extra electrons, in order to complete it's octet.

In other molecules or ions, the central atom may have more than an octet of electrons around it. This will occur *only where the central atom has an atomic number greater than ten.* Thus nitrogen (atomic number 7) will never have more than an octet of electrons; however, phosphorus (atomic number 15), which is directly below nitrogen in the periodic chart, can have more than an octet of electrons. Thus, elements with atomic numbers greater than ten can sometimes have either ten electrons (five pairs) or twelve electrons (six pairs) around them. For example SF_4 will have a dot structure

$$\ddot{F} \cdot \ddot{S} \cdot \ddot{F} \quad \text{, or} \quad F-\ddot{S}-F \text{ , where S has four bonding pairs and one lone}$$

pair of electrons around it — a total of five pairs (ten) electrons. Similarly another sulfur fluoride has the formula SF_6. In this case S has a total of six bonding pairs (or twelve) electrons around it.

$$\ddot{F}\colon\ddot{F} \quad \text{, or} \quad F \quad F$$

An Alternate Approach: When all atoms in a molecule or ion are expected to have an octet of electrons (which means that this approach will exclude hydrogen containing species), a simple formula may be used to determine the number of bonds (electron pairs) that are required.

The number of bonds
in molecule or ion
$$= \frac{[(\text{# of atoms}) \times (8)] - [\text{Sum of Group # of all atoms}]}{2}$$

Thus, in the examples above, for CCl_4, the equation becomes

The number of bonds = {[(5 atoms) × (8)] – [IV (for C) + 4 × VII (for Cl's)]}/2
The number of bonds = [40 – (4 + 28)]/2 = 4

Note: for an ion, the equation is modified to account for the charge, where the number of bonds is equal to

$$\frac{[(\text{# of atoms}) \times (8)] - [\text{Sum of Group # of all atoms}] - [\text{Charge of ion}]}{2}$$

Example: draw the dot structure for BF_4^-.

Instead of beginning with the dot structure, and attempting to assign all species an octet, we will instead determine the number of bonds first, using the formula given above.

$$\text{Number of bonds} = \frac{\{[(5 \text{ atoms}) \times (8)] - [\text{III (for B)} + 4 \times \text{VII (for F)}] - (-1)\}}{2}$$

$$\text{Number of bonds} = \frac{40 - [3 + 28] + 1}{2} = 4$$

The results of the calculation are that BF_4^- should have four bonds. Indeed a Lewis dot structure draw according to the instructions above result in an identical result. Thus,

1. Boron is the unique atom, and is assumed to be central. It is in Group III in the periodic chart, so we have $\cdot\ddot{B}\cdot$
   However, the negative charge on BF_4^- is assigned to the central B atom, so the initial dot structure becomes $\cdot\ddot{B}\cdot$

169

2. The four F atoms (Group VII) with their seven electrons are placed around the **B**

3. Each F atom is one electron shy of an octet and the B atom is four electrons shy of it's octet. The formation of four bonds, as indicated by the equation, will result in all atoms completing an octet of electrons. The structure is then

, which is the same as

Procedure

   VSEPR Theory states that the shape of a molecule or ion is determined by the number of pairs of electrons around the central atom, where the structure will minimize the repulsive interaction between pairs of electrons (irrespective of whether those pairs of electrons are bonding pairs or lone pairs). Simply stated, the pairs of electrons "get as far apart as possible". You will use ballons to represent electron pairs. The number of pairs of electrons will be the same as the number of ballons. You will make ballon models to verify the structures.

1. Inflate three different colored, but otherwise identical, ballons to the same size. (The ballons which will work the best should be "fat" and the same size at both ends, i.e. an ellipsoid). Twist each ballon in half, making the twist several full turns. Your three ballons should now resemble a figure eight. Now take two of your "figure eight" ballons and securely twist those two ballons together. Take the third "figure eight" ballon and twist it together with the other two. The resultant structure should have six lobes. Record your observations and answer the questions. Now pop one lobe on one of the ballons, which will leave five lobes. Record your results and answer the questions. Pop the other lobe of the ballon which you popped previously (the same color). Again, record your observations and answer the questions about this four-lobed structure. Pop another lobe, and answer the questions about this three-lobed structure. Finally pop the other lobe of the same ballon which you justed popped, leaving only two lobes left.

2. To determine the structure of a species using VSEPR Theory: 1) draw the Lewis dot structure, 2) count the total number of electron pairs around the central atom, and 3) assign the structure. VSEPR Theory states that the shape of an atom or molecule is determined by total number of electron pairs around the central atom. The total number of electron pairs includes both bonding pairs and lone pairs. However, we cannot "see" electrons or electron pairs; therefore lone pairs of electrons effect the shape of the species, whether we can "see" them or not. It is convenient therefore when determining the shape of a molecule or ion that we distinguish between the electron pair structure (which includes any lone pairs of electrons) and the geometry of the species (which includes only atoms).

For example, in determining the structure for the water molecule H_2O, first we draw the Lewis dot structure, as described above, to obtain

$$H-\overset{..}{\underset{|}{O}}: \qquad = \qquad H-\overset{..}{\underset{|}{O}}-H \qquad = \qquad :\overset{..}{\underset{|}{O}}-H$$
$$\quad\; H \qquad\qquad\qquad\qquad\qquad\qquad\qquad H$$

(Note: you should convince yourself that projecting a three-dimensional structure such as a tetrahedron onto a two-dimensional surface, i.e. the surface of the paper, means that all of three of the dot structures of H_2O are the same.) The O atom has a total of *four pairs* of electrons around it. Therefore the electron pair geometry (which determines the shape) is *tetrahedral*. However, of the four pairs of electrons, *two* are *bonding pairs* and *two* are *lone pairs* of electrons. Since we can see only atoms and not electrons or electron lone pairs, the geometry of H_2O is simply described as *bent*.

By using some molecular models, or by reconstructing the structures from new ballons, you should complete the table on the answer sheet.

3. VSEPR Theory treats multiple bonds as if they were a single pair of electrons when determining the molecular geometry. The rationale is that since a multiple bond between two atoms occupies the same general space as a single bond, then a multiple simply counts as one pair (one ballon) as far as molecular geometry is concerned. Thus, as applied to VSEPR Theory, the phrase "the total number of pairs of electrons around the central atom" means the number of bonding electron pairs (where a multiple bond counts as only one pair) plus the number of lone pairs.

For example, to determine the structure of CO_2, we first determine that the dot structure is

$$:\overset{..}{O}: \; :C: \; :\overset{..}{O}: \; = \; O{=}C{=}O$$

according to the rules described above. The carbon atom has two double bonds around it, but they only count as a total of *two* pairs of electrons. The electron pair structure, as well as the molecular geometry, is *linear*.

Complete the answer sheet for this section.

4. Using the information which you have worked out in the sections above, complete the answer sheet for this section.

Experiment 28
Preparation of Sodium Bicarbonate and Sodium Carbonate

Object

Sodium bicarbonate (sodium hydrogen carbonate) will be synthesized and then be used to prepare sodium carbonate.

Background

Sodium bicarbonate, $NaHCO_3$, and sodium carbonate, Na_2CO_3, are both important industrial chemicals and are thus made and used on a large scale. In addition to the familiar uses of sodium bicarbonate in baking and as an antacid, it is also used to prepare sodium carbonate. Sodium carbonate is used in cleaning powders, in the treatment of "hard" water, in the manufacture of glass, and as an inexpensive, convenient, industrial base chemical. Sodium bicarbonate and sodium carbonate are made by the Solvay process. A concentrated solution of ammonia in which sodium chloride is dissolved will absorb carbon dioxide.

$$NH_3 \; + \; CO_2 \xrightarrow[\;H_2O\;]{\;NaCl\;} NH_4^+ \; + \; HCO_3^-$$

Ammonium bicarbonate, NH_4HCO_3, is soluble in water, but sodium bicarbonate is not very soluble. Therefore, as bicarbonate anions are produced, they precipitate with the sodium cations (present from the dissolved NaCl) as sodium bicarbonate.

$$Na^+ \; + \; HCO_3^- \longrightarrow NaHCO_3 \downarrow$$

Reasonably pure sodium bicarbonate can be obtained by separating the precipitate by filtration, followed by drying at 100°C.

When sodium bicarbonate is heated to about 270°C, it decomposes to sodium carbonate, releasing water and carbon dioxide.

$$2\,NaHCO_3 \xrightarrow[\;270°\;]{\;\Delta\;} Na_2CO_3 \; + \; H_2O \; + \; CO_2 \uparrow$$

Sodium carbonate is a basic chemical because when dissolved in water it produces the hydroxide anion, OH^-.

$$Na_2CO_3 \; + \; H_2O \rightleftharpoons 2\,Na^+ \; + \; HCO_3^- \; + \; OH^-$$

Experimental Procedure

1. *Under the hood,* add 60 ml of concentrated ammonium hydroxide, NH_4OH or ammonia water, to a 125-ml flask. CAUTION: CONCENTRATED AMMONIUM HYDROXIDE SHOULD BE USED ONLY UNDER THE HOOD.

2. To the ammonium hydroxide, add in small portions quantities about 13 g of finely crushed sodium chloride, swirling after each addition to dissolve.

3. When no more sodium chloride will dissolve, that is, when the solution is saturated, carefully decant or filter the saturated solution into a 250 ml beaker. Be careful not to allow any undissolved sodium chloride crystals to transfer into the beaker. USE THE HOOD.

4. Add 25 to 50 g of dry ice (solid carbon dioxide) in the form of several pieces to the solution, using either tweezers or tongs to handle the dry ice. As the dry ice dissolves, sodium bicarbonate forms and precipitates as a white solid. Wait until all of the dry ice has disappeared (reacts or sublimes away).

5. Filter the precipitate through a suction filter apparatus, catching the sodium bicarbonate on the filter paper.

6. Dry the precipitate by placing it on a weighed watch glass placed over a beaker of gently boiling water. Heat it until the odor of ammonia can no longer be detected. Weigh the dried product.

7. Take about 1 g of your sodium bicarbonate and place it in a test tube. Heat the tube gently in a burner flame. Progressively heat the tube hotter as the compound "shrinks" due to the expulsion of carbon dioxide and water. When no further reaction is noted, set the tube on the lab desk to cool for at least 10 min.

8. Take some of the sodium carbonate from the test tube and dissolve it in 5 ml of H_2O. Test the solution with litmus paper and phenolphthalein solution. Describe the results.

9. Add a few drops of dilute sulfuric acid to the sodium carbonate solution; recording your results.

10. Repeat steps 8 and 9 with a solution of sodium bicarbonate.

Name_____   Lab Section_____

Station_____   Date_____

Preparation of Sodium Bicarbonate and Sodium Carbonate

Preparation

Weight of watch glass   _____ g

Weight of watch glass plus sodium bicarbonate   _____ g

Weight of sodium bicarbonate   _____ g

Why should you be careful in Step 3 that no sodium chloride is transferred?

What is the purpose of the dry ice?

Why is the sodium bicarbonate dried over boiling water?

What is the theoretical yield of sodium carbonate from 1.00 g of sodium bicarbonate? Show your work.

_____ g

Reactions

Describe the results of the test of sodium carbonate solution with litmus paper. Explain the results.

With phenolphthalein solution.

With H_2SO_4:

Describe the results of the test of sodium bicarbonate solution with litmus paper. Explain the results:

With phenolphthalein solution:

With H_2SO_4:

What would be the effect on the three previous questions if not all of the ammonia were driven off when drying the sodium bicarbonate?

Experiment 29

Preparation of an Inorganic Copper Compound

Object

An inorganic copper compound, called a coordination complex, will be prepared and partially characterized.

Background

There is a large class of compounds that contain a metal atom or ion bonded to several other molecules or ions by means of *coordinate* bonds. The complex compounds are called *coordination compounds*. The compound that will be prepared is tetraamminecopper(II) sulfate. Copper has an oxidation state of +2 and is bonded to four ammonia molecules by electron pair bonds made by sharing two electrons from each ammonia molecule with the copper ion. The ammonia molecules are called *ligands*.

$$Cu^{2+} + 4H_3N\colon \longrightarrow \begin{bmatrix} NH_3 \\ H_3N\colon \overset{\cdot\cdot}{Cu}\colon NH_3 \\ NH_3 \end{bmatrix}^{2+}$$

Tetramminecopper (II) ion

Coordination compounds are extremely important and common. For example hemoglobin is an iron(III) coordination complex, and chlorophyll is a magnesium coordination complex.

The preparation that is to be done in the experiment can be expressed as:

$$CuSO_4 + 4NH_3(\text{in excess}) \xrightarrow{H_2O} [Cu(NH_3)_4]SO_4 \cdot H_2O$$

Mol. wt.   Mol. wt.
159.6 amu   245.7 amu

If you started the experiment with 159.6 g of $CuSO_4$, you can, in theory, produce 245.7 g of product; that is, the theoretical yield of $[Cu(NH_3)_4]SO_4\cdot H_2O$ is 245.7 g. If you began with 79.8 g of $CuSO_4$ (one-half of 159.6), the theoretical yield is one-half of 245.7 g, or 122.8 g. If your *actual* yield (starting with 79.8 g of $CuSO_4$) was, say, 61.4 g of product, your yield would be 50% (or half of the theoretical yield).

Experimental Procedure

1. Dissolve approximately 5 g (weighed to the nearest 0.001 g) of copper sulfate, $CuSO_4$, in 70 ml of distilled water. The $CuSO_4$ should be added slowly with stirring with a glass stirring rod. Note and

record the color. The color is due to a coordination compound where the water molecules are acting as ligands.

2. *Under the hood slowly* add 25 ml [5mL] of concentrated ammonia (ammonium hydroxide), stirring the solution continuously. Note and record the color. The first addition of ammonia forms $Cu(OH)_2$. Further addition of ammonia forms the coordination complex $[Cu(NH_3)_4]^{2+}$.

3. Add 90 ml [15 mL] of ethyl alcohol (ethanol) to the solution with stirring. The precipitate (solid) formed is $[Cu(NH_3)_4]SO_4 \cdot H_2O$.

4. Allow the solution to stand for about 5 min, and then filter with a vacuum filter apparatus using a Büchner funnel. Scrape all of the solid from the beaker onto the filter paper. Pour two 10-ml [2 ml] portions of ethyl alcohol over the crystals, and let air pass over the crystals until they appear dry.

5. Weigh the dried product in a preweighed 125-ml conical or Erlenmeyer flask. Show the product to your instructor.

6. Weigh about 0.10 g [scoop] of the dried product that you have just prepared and divide it approximately in half. Place each half into a test tube. Half-fill one of the test tubes with water, and the other test tube half-full with ethyl alcohol. Shake both tubes to ensure good mixing. Record your observations concerning the solubility of the complex in water and in alcohol. [same time]

7. Calculate the theoretical and actual yield of the complex.

182

Experiment 30
Recovery of Aluminum

Object

Aluminum scrap metal will be converted by chemical processes into a useful compound.

Background

In many instances it is possible to transform scrap materials such as aluminum into new compounds that can be extremely useful. The new compounds may have a totally different appearance than the original scrap material, and they may be equally or more useful. The widespread use of aluminum for beverage cans has led to ecological problems because aluminum is much more resistant than steel to environmental degradation. Although aluminum is initially easily oxidized to Al_2O_3, this oxide covers the surface of the aluminum metal and protects the remaining aluminum from further oxidation. Although some companies have found that it is economically feasible to recycle aluminum cans, today's experiment demonstrates an alternative solution to the disposition of waste aluminum cans: the synthesis of a useful aluminum compound from such scrap metal.

Aluminum rapidly reacts with a hot aqueous (water) solution of potassium hydroxide.

$$2Al + 2KOH + 6H_2O \longrightarrow 2KAl(OH)_4 + 3H_2 \uparrow$$

Once the aluminum metal is converted into aluminum ions in solution, a number of different compounds can be easily made that are either useful in their own right or valuable as intermediates in the preparation of still other aluminum compounds.

Potassium aluminum sulfate $[KAl(SO_4)_2 \cdot 12H_2O]$, or "alum" is an aluminum compound with many valuable uses ranging from the manufacture of pickles to the dyeing of fabrics. Alum is a "double salt", yielding the ions K^+, $[Al(H_2O)_6]^{3+}$, and SO_4^{2-} when dissolved in water. Alum is made from $KAl(OH)_4$ by the addition of sulfuric acid

$$2KAl(OH)_4 + H_2SO_4 \longrightarrow 2Al(OH)_3 + K_2SO_4 + 2H_2O$$

$$2Al(OH)_3 + 3H_2SO_4 \longrightarrow 2Al^{3+} + 3SO_4^{2-} + 6H_2O$$

$$\text{Cooling of solution} \longrightarrow [KAl(SO_4)_2 \cdot 12H_2O] \text{ crystals}$$

Experimental Procedure

1. Weigh 1 g (±0.01 g) of aluminum scrap, cut into *very* small pieces, and put these in a 250-ml flask. The smaller the pieces of aluminum, the more rapid the reaction will be. Aluminum foil may be used in place of the pieces of scrap aluminum.

2. Add 50 ml of potassium hydroxide, KOH, to the aluminum. CAUTION: WEAR SAFETY GLASSES. BE CAREFUL! DO NOT SPLATTER THE SOLUTION. KOH IS CAUSTIC.

185

3. Heat the beaker *gently* over a small flame. Hydrogen will be evolved, so this step should be performed in a well ventilated area. Heating should be continued until all the aluminum is reacted. Some impurities may remain. During the heating if the liquid level drops below one-fourth the original level, add water to bring the level back to one-half the original liquid level. This process may have to be repeated several times during the reaction of the aluminum.

4. When all of the aluminum has reacted, filter the warm solution through a *thin* layer of glass wool in a funnel.

5. Allow the clear solution to cool, stirring occasionally. Slowly add 30 ml of 6 M H_2SO_4, stirring continuously. The solution will contain large lumps of aluminum hydroxide after all of the H_2SO_4 has been added.

6. The reaction mixture should be gently heated again for about 10 min. or until all of the aluminum hydroxide dissolves.

7. After the solution is clear, remove it from the heat, and filter (only if any solids are present). Cool the beaker containing the solution in an ice bath for about 20 min. Alum crystals should be present in the beaker.

8. Suction filter the crystals using a Buchner flask. Wash the crystals (which should be on the filter) with 20 ml of a 50/50 alcohol-water mixture (in which the alum is not very soluble) by pouring the alcohol-water mixtures over the crystals while they are on the filter paper. Suck the crystals dry.

9. Weigh the dried crystals of alum and show it to your teaching assistant.

10. Determine the melting point of your crystalline alum. This is done by placing finely ground alum into the bottom of a melting point capillary tube to a depth of about 0.5 cm. Place a small rubber band about 5 cm above the bulb or bottom end of the thermometer. Insert the capillary tube containing the alum under the rubber band with the closed end of the capillary even with the thermometer bulb. Place the thermometer into a cork or rubber stopper so that it may be attached to a buret clamp. Place the thermometer with the attached capillary tube in a 250-ml beaker containing enough water to cover the thermometer bulb, making sure that the open end of the capillary is above the water level. See Figure 30-I. Heat the water slowly so that the water temperature increases about 3° per min. Carefully watch the solid in the capillary tube. At the moment the solid melts, note the temperature. This is the melting point of the alum.

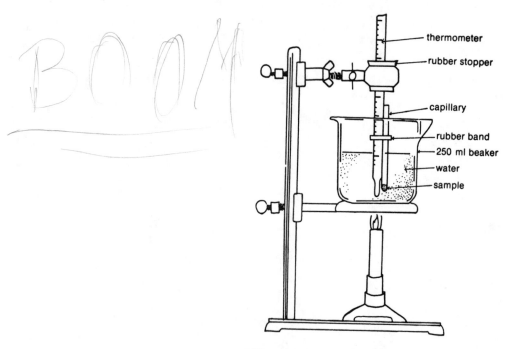

thermometer
rubber stopper
capillary
rubber band
250 ml beaker
water
sample

Figure 30-I. Melting point apparatus

$$\frac{0.5 \text{ g Al}}{} \left| \frac{1 \text{ mol Al}}{27 \text{ g Al}} \right| \frac{1 \text{ mol } [\text{KAl}(SO_4)_2 \cdot 12 H_2O]}{1 \text{ mol Al}} \left| \frac{474 \text{ g Al}(SO_4)_2}{1 \text{ mol } \text{KAl}(SO_4)_2} \right.$$

$$\frac{.5 \cdot 474}{27} \qquad \frac{474}{54}$$

187

Experiment 31
The Synthesis of the Drug Aspirin

Object

To synthesize a common drug by an organic reaction known as esterification.

Background

One of the most satisfactory analgesic (pain relieving) and antipyretic (fever reducing) drugs is acetylsalicylic acid, commonly known as aspirin. Its use can be traced back many centuries to the use of willow bark in treating fever and pain. The active ingredient in willow bark is salicylic acid, which has the disadvantages of being extremely irritating to the mouth and stomach, as well as being very disagreeable to taste. In 1899 a derivative of salicylic acid, acetylsalicylic acid, was introduced into medicine. It is still regarded today as being as effective as it was when it was first introduced. Aspirin represents one of the most common drugs in use, with over two million pounds being produced annually in the United States.

The preparation of aspirin demonstrates many of the techniques involved in the synthesis of a simple organic compound. Aspirin is made by the reaction of salicylic acid with acetic anhydride, using sulfuric acid as a catalyst.

Salicylic Acid	Acetic Anhydride	Acetylsalicylic Acid	Acetic Acid
Mol. Wt. 138	Mol. Wt. 102	Mol. Wt. 180	Mol. Wt. 60

Experimental Procedure

1. Accurately weigh 6 g (±0.01 g) of salicylic acid in a 125-ml Erlenmeyer flask. Salicylic acid is an odorless, white, crystalline solid. Wash off any acid that comes in contact with the skin, as it is irritating to the skin and nasal passages.

2. Obtain 8 ml (density 1.08 g/ml) of acetic anhydride and add it to the salicylic acid in the flask under the hood. CAUTION: ACETIC ANHYDRIDE IS IRRITATING TO THE SKIN. ANY THAT IS SPILLED ON THE SKIN SHOULD BE WASHED OFF IMMEDIATELY WITH A LARGE VOLUME OF WATER. (Note: Acetic anhydride will react with water to form acetic acid).

3. Add 10 drops of concentrated sulfuric acid, H_2SO_4 to the mixture. Immediately swirl the flask gently, and heat the flask in beaker of boiling water (Figure 31-I) for 15 min. CAUTION: SULFURIC ACID IS VERY DANGEROUS. IT REACTS VIOLENTLY WITH WATER, IS VERY IRRITATING TO THE SKIN, AND IS DESTRUCTIVE TO CLOTHING. USE SAFETY GLASSES. If after 15 minutes, all of the

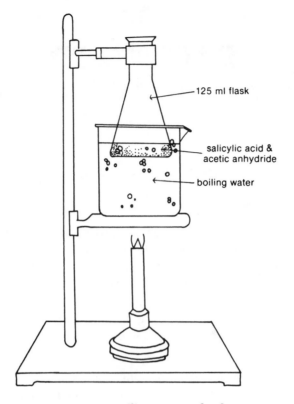

125 ml flask

salicylic acid & acetic anhydride

boiling water

Figure 31-I. Boiling water bath

salicylic acid is not dissolved, heat for an additional 10 min. Some stirring may be necessary. Failure to get all of the salicylic acid to dissolve will result in a poor yield.

4. Remove the flask from the boiling water bath, wait 10 min, and carefully add 25 ml of ice water to the flask. Set the flask in a beaker of ice until the crystallization of the aspirin appears complete (about 10 min.).

5. Separate the aspirin crystals from the liquid by suction filtration using a Büchner funnel. Discard the liquid.

6. Recrystallize the aspirin by dissolving the crystals in 20 ml of ethyl alcohol (ethanol) in a 100-ml beaker. If it is necessary in order to get the crystals to dissolve, heat the beaker containing the alcohol and the aspirin in a water bath (as in step 3, Figure 31-I). Then pour 50 ml of warm distilled water into the solution. Set the beaker containing the aspirin, alcohol, and water aside to cool. The beaker may be set in a beaker of ice to speed cooling.

7. After recrystallization is complete, suction filter the aspirin again using the Büchner funnel. Dry the aspirin by spreading the crystals on a dry piece of filter paper and then by patting them dry with another piece of filter paper.

8. Weigh the aspirin in a preweighed small flask or beaker.

9. Calculate the percent yield.

10. Show the dried aspirin crystals to your instructor, and allow him to comment on their appearance on your answer sheet.

Name_____   Lab Section_____ ____

Station_____   Date _____ ____

The Synthesis of the Drug Aspirin

Observations

Weight of the 125-ml flask   _____

Weight of the 125-m flask plus salicylic acid   _____ _____

Weight of the salicylic acid   _____

Weight of empty flask or beaker (in which the aspirin is
to be weighed.)   _____

Weight of flask plus aspirin   _____

Weight of aspirin   _____ _____

Calculation of the theoretical yield of aspirin (Show your work).

                                                            _____

Calculation of the percent yield (Show your work).

                                                            _____ %

Why is the amount of acetic anhydride used in the experiment not important in calculating the
theoretical yield of aspirin?

List possible reasons why your yield of aspirin wasn't 100%?

Instructors comments:

Are there any likely experimental errors that would result in your finding greater than a 100% yield? If so, what are they?

What is the purpose of the recrystallization?

Why is water added to the alcohol during recrystallization?

Experiment 32
Soap — Preparation and Properties

Object

A soap will be prepared by a simple reaction. Some of the simple chemical and physical properties will be investigated.

Background

The sodium or potassium salt of a long-chain fatty acid (a type of organic compound) is known as a soap. Solid soaps are usually sodium salts. Such soaps are simply made by the reaction of a fat or oil (which are glycerol esters of the fatty acid) such as cooking oil or cottonseed oil with concentrated sodium hydroxide. Ethyl alcohol is added as a common solvent for the reactants so as to speed up the reaction, although it is not a necessary ingredient.

The saponification, or soap making, reaction is

$$
\begin{array}{ccccc}
\underset{\text{O}}{\overset{\text{O}}{\parallel}} & & & & \\
\text{CH}_2\text{-O-C-R}_1 & & \text{CH}_2\text{OH} & & \text{R}_1\text{C-O}^-\text{Na}^+ \\
\underset{\text{O}}{\overset{\text{O}}{\parallel}} & & & & \underset{\text{O}}{\overset{\text{O}}{\parallel}} \\
\text{CH-O-C-R}_2 & + \quad 3\text{NaOH} \rightarrow & \text{CHOH} & + & \text{R}_2\text{C-O}^-\text{Na}^+ \\
\underset{\text{O}}{\overset{\text{O}}{\parallel}} & & & & \underset{\text{O}}{\overset{\text{O}}{\parallel}} \\
\text{CH}_2\text{-O-C-R}_3 & & \text{CH}_2\text{OH} & & \text{R}_3\text{C-O}^-\text{Na}^+
\end{array}
$$

An oil (a glyceride or ester   Glycerol   Soap — Sodium salts of
of fatty acids and glycerol)   the fatty acids

Soaps have a polar end of the molecule that is hydrophilic (water-loving) and a long nonpolar organic "chain" that is hydrophobic (water-hating). Soap is able to dissolve (actually suspend or emulsify) oils or other organic materials in water in the following way: the polar end of the soap molecule dissolves in water while the nonpolar end incases or surrounds the small globules of oil or "dirt", effectively making the oil "soluble" in water. If the water in which the soap is dissolved contains appreciable amounts of such ions as calcium, magnesium, or iron (that is, the water is "hard"), then these ions replace the sodium in the soap molecule, and the resulting compound becomes insoluble, forming "bath tub ring" or scum.

Experimental Procedure

1. In a 150-ml beaker weigh out 20 g of vegetable oil or lard (whichever is furnished).

2. Add 20 ml of ethyl alcohol (ethanol) and 25 ml of 20% sodium hydroxide solution to the oil, stirring the resulting mixture.

3. Place the beaker on a wire gauze supported on a ring and stand and heat *gently*. PRECAUTION: KEEP THE FLAME AWAY FROM THE TOP OF THE BEAKER, OR THE ALCOHOL WILL IGNITE, RESULTING IN SOAP FLAME. If the alcohol does ignite, simply cover the beaker with a watch glass or piece of asbestos to extinguish the flame. Continue heating until the odor of alcohol disappears and pasty mass remains. This pasty mass is the soap plus the glycerol produced in the reaction. Set the beaker on the lab desk to cool.

4. Add 100 ml of a saturated solution of sodium chloride to your soap mixture and stir thoroughly with a glass rod. This process is referred to as "salting out" and is used to remove the soap from water, glycerol, and unreacted sodium hydroxide.

5. Filter the soap on a suction funnel and wash the soap once with a small quantity of ice water while the suction is still on.

6. Weigh your dried soap and show the soap to your instructor.

7. Wash your hands with a small amount of your soap using distilled water. If too much oil was used in the preparation, the soap will feel greasy. Too much sodium hydroxide will also result in a slick feeling, and will result in a roughening of your hands. Record your observations as indicated.

8. Place about 10 drops of kerosine in a test tube containing 10 ml of water. Shake (the test tube). What happens? Let stand for a few minutes. What happens? Prepare another test tube as above, but also add a small amount of your soap. Shake the tube as above and let it stand. Record your observations.

9. Take about 1 g of your soap and warm it with 50 ml of distilled water in a 100-ml beaker. When the solution is reasonably clear, pour 15 ml into each of three test tubes. Add some 5% $CaCl_2$ solution to one test tube, some 5% $MgCl_2$ to another test tube, and some 5% $FeCl_3$ to the third test tube. Describe the results on your answer sheet.

10. Soap with free alkali (base) can be damaging to your skin. To test to see if your soap contains free alkali, dissolve a small piece in 15 ml of ethyl alcohol and then add two drops of phenolphthalein. Record what happens.

Name _____   Lab Section_____

Station_____   Date _____

Soap — Preparation and Properties

Preparation

Weight of soap   _____

What property of ethyl alcohol makes it useful in the in the preparation
of soap?

Instructor observations:

Properties

Washing properties: How did your soap feel when used for washing. What does this indicate?

Emulsification of oil: Oil-water mixture. Describe and explain the results of this experiment.

: Oil-water-soap mixture. Describe and explain the results of this experiment.
Explain any differences between this results as compared to those above without the use of soap.

Hard water reactions: What happened when $CaCl_2$ was added to the soap solution? Why?

: $MgCl_2$?

: $FeCl_3$?

Basicity of soap: Record your observations when phenolphthalein was added to a soap solution. Explain the results.

Experiment 33

The Analysis of Solids in Cigarette Smoke

Object

In this experiment you will determine the amount of solids in cigarette smoke and evaluate the effectiveness of cigarette filters.

Background

The average person breathes about 35 pounds of air each day to provide oxygen to our blood, which uses oxygen to perform metabolic processes essential to life. The air that we breathe is usually not pure, but rather is filled with pollutants which shortens our lives. Polluted air is a major factor in emphysema, which kills about 50,000 people in the United States each year, and in chronic bronchitis, which affects about one out of five men between the ages of 40 and 60. Yet with all of the dirty air that we are forced to breathe, millions add to the problem by creating additional air pollution, personal air pollution, by cigarette smoking. The Surgeon General of the United States has published research proving that smoking is a direct cause of lung cancer, and a warning to that effect is required on all cigarette packages and advertisements. The death rate from lung cancer is about ten times higher for cigarette smokers than for nonsmokers. Cigarette smoke irritates the bronchial tubes and renders them much more susceptible to disease. Some of the gases in cigarette smoke are carcenogenic. There is also strong evidence linking smoking with heart disease.

Solids and small particles are a major component of cigarette smoke. The smoker inhales these and they can pass through the body's defense mechanisms and destroy lung tissue. These particles range in size from submicron to several microns (10^{-6} meters), and they are in part responsible for emphysema and chronic bronchitis. Some of these particles are mineral dust (such as lead), fly ash, and organic tars. In the lung they can irritate and destroy the alveoli (tiny sacs in the lung in which gas exchange occurs) and thus cause emphysema.

In the first part of the experiment you will determine the amount of solids in cigarette smoke. Tested will be: (a) a non-filter cigarette, (b) a filter-tip cigarette, (c) a filter-tip cigarette with the filter removed, and (d) a "little cigar". In the second part of the experiment you will test the effectiveness of the filter in a filter-tip cigarette. This will be done by first determining the amount of solids that get through the filter when one-third of the cigarette is burned. This will be compared to the amount of solids that get through the filter when the cigarette is burned another one-third, and again to the solids passing through the filter when the last third of the cigarette is burned.

Experimental Procedure

1. Obtain a 250-ml suction flask, a piece of filter paper, a one-hole rubber stopper, some glass tubing, and a piece of rubber tubing. Assemble the equipment as shown in Figure 33-I, and attach the suction flask to a water aspirator.

2. Remove the filter paper from the apparatus and weigh it. All weighings should be measured to the nearest 0.001 g. Return the filter paper to the apparatus.

3. Obtain a non-filter-tip cigarette and weigh it. Then insert the cigarette into the glass cigarette holder.

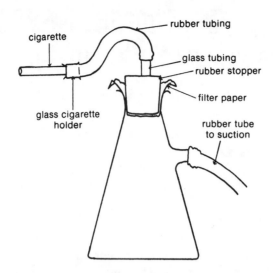

Figure 33-I. Cigarette smoke set up

4. Turn on the water aspirator and light the cigarette. It is very important in order to obtain good results that the aspirator be adjusted so that the cigarette burns slowly and evenly over a period of 3 to 5 minutes. If the burning proceeds too rapidly, you will obtain poor experimental results.

5. After the cigarette has been almost totally consumed, or down to the "butt", turn off the aspirator, and extinguish the cigarette with a single drop of water. Remove the filter paper from the apparatus and weigh it.

6. Weigh the portion of the cigarette that was not consumed. Subtract this weight from the initial weight of the cigarette. Calculate the number of milligrams of solids collected per gram of cigarette consumed.

7. Repeat the above procedure with each of the following: a filter-tip cigarette, a filter-tip cigarette with the filter removed, and a "little cigar"

8. For the second part of this experiment for this week, obtain a filter-tip cigarette and, using a pencil or pen, mark off the length of the tobacco part of the cigarette into thirds.

9. Place the cigarette in the apparatus, using the same procedure that you used earlier. Also place a fresh piece of weighed filter paper in the apparatus.

10. Turn on the aspirator and light the cigarette. Allow the cigarette to burn slowly until one-third of the tobacco has been consumed. Turn off the aspirator, and extinguish the cigarette with one drop of water.

11. Weigh the filter paper and record the weight. Return the filter paper to the apparatus.

12. Turn on the aspirator, light the cigarette, and allow another third of the cigarette to burn. Turn off the aspirator, extinguish the cigarette with one drop of water. Remove the filter paper and weigh it, recording the weight.

13. Repeat the procedure with the last third of the cigarette.

200

Name_____   Lab Section_____   _____

Station_____   Date_____

Analysis of Solids in Cigarette Smoke

	Non-filter-tip cigarette	Filter-tip cigarette	Filter-tip cigarette minus filter	"Little cigar"
1. Initial wt. of filter paper	_____	_____	_____	_____
2. Weight of filter paper after the cigarette is consumed	_____	_____	_____	_____
3. Weight of the solids on the filter paper (mg)	_____	_____	_____	_____
4. Initial wt. of cigarette	_____	_____	_____	_____
5. Weight of the cigarette after burning	_____	_____	_____	_____
6. Weight of cigarette consumed	_____	_____	_____	_____
7. Weight of solids collected per g of cigarette consumed (mg/g)	_____	_____	_____	_____

Which of the cigarettes produced the most solids? Can you offer an explanation of your results?

Effectiveness of Cigarette Filters

1. Initial wt. of filter paper   _____

2. Wt. of filter paper after one-third of the
   cigarette is consumed   _____

3. Wt. of solids collected from the first third
   of the cigarette   _____

4. Wt. of filter paper after two-thirds of the
   cigarette is consumed   _____

5. Wt. of solids collected from the second third
   of the cigarette   _____

6. Wt. of filter paper after last third of the
   cigarette is consumed   _____

7. Wt. of solids collected from the last third of
   of the cigarette   _____

Which portion of the filter-tip cigarette yielded the most solids? Did you expect this result?
Explain.

Do you feel that smoking a filter-tip cigarette is much better than smoking a non-filter-tip cigarette?
Defend your answer on the basis of your experimental data.

What are some of the sources of experimental error in this experiment?

Experiment 34
Ethyl Alcohol — Preparation by Fermentation

Object

This experiment demonstrates the preparation of ethyl alcohol (ethanol) by fermentation of a sugar. The process of distillation will be used as a purification process.

Background

The action of certain microorganisms on compounds can result in the synthesis of many new compounds. One of the oldest and most widely used of such processes is the preparation of the ethyl alcohol by the action of yeast on certain sugars (carbohydrates).

$$C_6H_{12}O_6 \xrightarrow{\text{yeast}} 2C_2H_5OH + 2CO_2 \uparrow$$

glucose   ethyl alcohol

As the sugar is converted into alcohol, the concentration of alcohol builds until it finally kills the yeast that made it. Normally death of the yeast occurs when the alcohol content reaches about 15%. During the fermentation process, carbon dioxide is expelled. The remaining mixture contains yeast, water, and alcohol. The alcohol can be removed from the mixture by distillation, since pure ethanol boils at about 78°, much lower than water. When a mixture of ethyl alcohol and water is heated, the vapor contains a much higher concentration of ethyl alcohol than water, and if the vapor is condensed, a liquid can be obtained that has a very high alcohol content. This is the process of distillation. The percentage of alcohol in the distillate (distilled material) can be estimated by measuring the density, as pure ethyl alcohol has a density of 0.79 g/ml and water has a density of 1.00 g/ml.

Ethyl alcohol undergoes many reactions. Usually the fermentation mixture contains some acetic acid and some acetaldehyde, both of which are formed by the partial oxidation of ethanol.

$$C_2H_5OH \xrightarrow{O_2} CH_3CHO \text{ (acetaldehyde)}$$

$$C_2H_5OH \xrightarrow{O_2} CH_3COOH \text{ (acetic acid)}$$

Complete oxidation of ethyl alcohol produces CO_2 and H_2O. In the presence of a base, ethyl alcohol is oxidized to iodoform, CH_3I, a yellow solid that has been used as an antiseptic.

$$C_2H_5OH + 4I_2 + 6NaOH \longrightarrow CH_3I + NaCOOH + 5NaI + 5H_2O$$

Experimental Procedure

1. First week. In a 500-ml Erlenmeyer flask, place 200 ml water, 30 ml of corn syrup, and about one-eight of a cake of yeast, and 12 ml of nutrient salt solution (The nutrient salt solution for the yeast is made by dissolving 2 g K_3PO_4, 0.2 g $Ca_3(PO_4)_2$, 0.2 g $MgSO_4$ and 10 g $NH_4C_2H_3O_2$ in 900 ml of water). Stuff a paper towel into the top of the flask, write your name on the flask, and store it for a week in an area indicated by your teaching assistant.

2. Second week. Take your fermentation mixture and decant the liquid into another flask to separate it from the residue. An alternative to the above instructions is to take 200 ml of freshly fermented corn syrup that is provided by your teaching assistant.

3. Set up a distillation apparatus as shown in Figure 34-I. (CAUTION: Be careful when inserting the thermometer through the rubber stopper. Undo pressure will break the thermometer, forcing

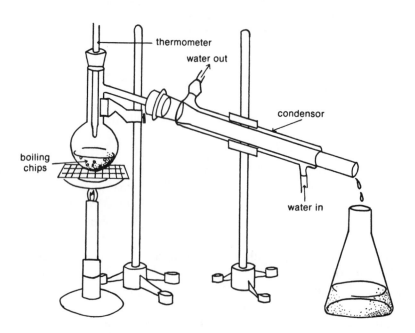

Figure 34-I. Distillation apparatus

the broken end into the palm of your hand.) Add your fermentation liquid to the distillation flask together with several boiling chips. Heat the mixture *very* slowly, bringing the liquid to a *slight simmer!* Heating too rapidly will cause the liquid to bump, forcing impure liquid into your receiving vessel. In order to insure against overheating, you may wish to place the distillation flask into a large beaker of water, and then heat the water.

4. Record the temperature at which the liquid begins to distill over. Distill over about 15 ml of liquid as slowly and gently as possible. When 15 ml has been collected, remove the receiving vessel, shut off the flame, and stop the distillation. Show the liquid to your teacher, and allow him to record his observations on your data sheet.

5. Weigh an empty 10-ml graduated cylinder. Transfer *exactly* 10 ml of your distilled alcohol into the cylinder and weigh it again. Determine the density.

6. Take several ml (no more than 5 ml) and note the odor, recording your observations on the data sheet. Pour some of the alcohol into an evaporating dish and set a match to it, recording your observations.

7. Take another portion of your distilled alcohol and add an equal quantity of water. Note whether the two liquids are miscible (whether they mix). Attempt to set fire to a small quantity of this mixture in an evaporating dish. Record all observations.

8. Take about 3 ml of your distilled alcohol and add some iodine crystals. Shake gently until a dark brown solution is produced. Then add, *dropwise,* sodium hydroxide solution. Describe in detail your observations.

Name _____   Lab Section _____

Section _____   Date _____

Ethyl Alcohol — Preparation by Fermentation

<u>Preparation</u>

What is a possible purpose of stuffing a paper towel into the flask during fermentation?

Describe the appearance of the flask after a week's fermentation.

What is the solid material at the bottom of the flask?

<u>Distillation</u>

Temperature at which liquid begins to distill   _____°C

Evaluation of distilled alcohol by teacher.

What function do the boiling chips serve?

<u>Density</u>

Weight of empty 10-ml graduated cylinder   _____

Weight of graduater cylinder plus 10-ml distilled alcohol   _____

Weight of distilled alcohol   _____

Density of the distilled alcohol   _____

What can you conclude from the density concerning the purity of your distilled alcohol?

Comment on the odor of pure distillate.

Flammability of distillate.

Comment on the flammability of the distillate.

Write a balanced equation for the burning of alcohol.

Alcohol-water mixture

Do the alcohol and water mix? _____

Record any observations.

Does the mixture burn? Explain.

Iodoform. Describe the iodoform reaction as completely as possible, offering explanations for your observations.

Experiment 35
Solubility of $CaSO_4$

Object

In this experiment, you will measure the solubility of calcium sulfate in water. The $CaSO_4$ solution will be passed through an ion exchange column and each Ca^{2+} ion will be replaced by two H^+ ions. Thus, the $CaSO_4$ solution is converted into a H_2SO_4 solution. You will measure the amount of H_2SO_4 in the solution by titrating it with NaOH to the phenolphthalein end point. You will also determine the charge on an unknown metal ion by a similar procedure.

Background

You will assume that a solution in contact with an excess of solid calcium sulfate for several weeks is a saturated solution. You will measure the amount of calcium sulfate dissolved in the solution by an ion exchange and titration procedure which is outlined here.

You will first convert the ion exchange resin (small plastic beads) in a column into the H^+ form. Simply passing H_2SO_4 solution through the column will accomplish this. You will then rinse the column well so that no acid is present in water which passes through the column. You will then pass an accurately measured volume of the $CaSO_4$ solution through the column. The chemical reaction in the column is described by the equation

$$2R^- H^+ \; + \; Ca^{2+}(soln) \longrightarrow (R^-)_2 Ca^{2+} \; + \; 2H^+(soln)$$

R^- stands for a sulfonic acid anion which is covalently bonded to the polymer. A cation will be attached to the R^- by an ionic bond. The reaction has the effect of removing Ca^{2+} ions from the solution and replacing each Ca^{2+} by two H^+ ions.

The H_2SO_4 solution which comes out of the column is titrated with NaOH in the usual manner to the phenolphthalein end point. Calculations of the $CaSO_4$ content of the saturated solution are based on the following:

$$V_{NaOH} \; M_{NaOH} \; = \; n_{NaOH} \; = \; n_{H^+}$$

n_{H^+} represents the number of moles of H^+ obtained from exchanging Ca^{2+} by two H^+ in the original solution.

$$\tfrac{1}{2}n_{H^+} \; = \; n_{Ca^{2+}}$$

$$\text{moles } Ca^{2+}/\text{liters of } CaSO_4 \; = \; M_{CaSO_4}$$

To obtain the moles of Ca^{2+} in the sample, the moles of H^+ must be multiplied by $\tfrac{1}{2}$. To obtain the concentration of $CaSO_4$ in the saturated solution, you divide the moles of Ca^{2+} in the sample by the number of liters in the sample. Since $CaSO_4$ is completely ionized in solution, the concentration of Ca^{2+} and the concentration of SO_4^{2-} will be the same as the moles of $CaSO_4$ per liter calculated above. The solubility product for the process

$$CaSO_4(s) \rightleftharpoons Ca^{2+} \; + \; SO_4^{2-} \quad \text{is} \quad K_{sp} \; = \; [Ca^{2+}][SO_4^{2-}],$$

Next you will determine the charge on a metal ion by a similar procedure. You will be given a 0.1M solution of the metal and will be asked to measure the charge on the metal ion and to observe the behavior of the metal ion on the ion exchange column. You will utilize the fact that, according to the following equation

$$nRH + X^{+n}(soln) \longrightarrow R_nX + nH^+(soln)$$

nH^+ ions will be produced in solution when a metal ion of charge $+n$ is ion exchanged. You will need to work out your own method for calculating the charge on the unknown metal ion.

Procedure

1. Obtain a column filled with ion exchange resin from the stockroom. If the resin contains bubbles, fill the column with distilled water, stopper the column, and invert it to permit the bubbles to collect. Turn the column upright and let the resin settle. Do not let the resin bed become dry at any time!

2. Convert the resin into the H^+ form as outlined here. Drain excess water from the bottom. Pour 25 ml of 1.5 M H_2SO_4 on the column; pass it through the column. Just before the solution level permits the resin to get dry, squirt in about 5 ml of distilled water. Repeat the addition of 5 ml portions of distilled water until ~30 ml of water have passed through the column. Collect ~10 ml more water, add one drop of phenolphthalein indicator and one drop of 0.1M NaOH. If the solution is basic (pink) then you have washed all the acid out of the column. If the solution is not pink, wash the column with another 20 ml of water and test again. When you have washed all the acid from the column, clamp the column to stop the flow, and go on to the next part.

3. Pipet 50 ml of saturated $CaSO_4$ solution into the column. Collect all liquid from the column in a 250 ml Erlenmeyer flask. You will have to empty the pipet into the column slowly. When the solution runs through the column, wash with 5 ml portions of distilled water, as before, until you have collected a total volume of 100 to 150 ml. Add two drops of phenolphthalein.

4. Set up a buret, clean and dry it, and fill it with NaOH solution. Record the molarity of NaOH from the bottle you use. Record the initial solution volume, and titrate the ion exchanged solution until the solution remains pink for 30 seconds. Remember to add NaOH very slowly during the last part of the titration. You are trying to only go one drop (or less) past the end point, and this can be done only by adding the last portion of NaOH to the flask one drop at a time. Record the final volume of NaOH in the buret.

5. Convert the resin into the H^+ form and rinse with distilled water as outlined in step 2.

6. Obtain 5 ml of the unknown salt solution in a 10 ml graduated cylinder. This solution is 0.1M in the metal ion, but the type of metal ion is unknown. Pour the solution into the column, rinse with several 5 ml portions of distilled water, and continue until a total volume of ~50 ml is collected. Add phenolphthalein, read the buret, titrate to the end point, and reread the buret.

7. Convert the resin into the H^+ form as before. Record your observations. There is no need to rinse all acid from the column at this point.

210

Name_____   Lab Section_____

Station Number_____   Date_____

Solubility of $CaSO_4$

Ion exchange column contains   _____ ml of resin.

Appearance of saturated $CaSO_4$ solution:

_____ ml or_____ l of $CaSO_4$ solution placed in ion exchange column.

_____ ml of solution after ion exchange.

Molarity of NaOH solution, _____

final buret reading _____

initial buret reading_____

volume of NaOH_____ ml or_____ l

Show your calculations below for moles of H^+ :

_____ moles of H^+,_____ moles of Ca^{2+} in sample concentrations of

$CaSO_4$ in sample_____

$[Ca^{2+}]$ = _____ $[SO_4{}^{2-}]$ = _____

$K_{sp} = [Ca^{2+}] [SO_4{}^{2-}]$ = _____ (experimental value)

211

Appearance of unknown 0.1 M salt solution:

Behavior of unknown salt solution on column:

Behavior of column when converting resin to H^+ form:

Appearance of final solution washed from column:

_____ml of unknown 0.1 M salt

final buret reading _____

initial buret reading _____

ml of NaOH _____

Show your calculations below:

Charge on metal ion is _____.

Experiment 36
The Solubility of a Salt in Water

Object

The solubility of a salt in water at various temperatures will be determined to see how solubility varies with temperature.

Background

When a salt dissolves in water, a homogeneous mixture is formed called a *solution*. The water is called the *solvent*, and the salt is called the *solute*. At a given temperature only a certain quantity of solute will dissolve in the solvent, and the solution is said to be *saturated*. The exact amount of solute that will dissolve in a solvent at a given temperature is referred to as the solubility of that solute in that solvent. The solubility is usually measured in grams of solute per 100 grams of solvent. For example, the solubility of KBr at 25°C is 64 g per 100 g of water. That means that at 25°C, exactly 64 g of KBr will dissolve in 100 g of water. The solution is saturated, and at 25°C no more KBr will dissolve. If excess solute is placed in a saturated solution (for example, more KBr is added to the saturated KBr-water solution), more solute *will* dissolve, but only at the same rate that some KBr "comes out" of solution, or crystallizes. Thus the net amount in solution is the same, and the system is said to be in dynamic equilibrium. The rate of solute dissolving equals the rate of solute crystallizing.

When saturated solutions of solid solutes in liquid solvents are cooled, the excess solute usually separates from the solution by recrystallizing. However, if at saturated solution is cooled undisturbed with no excess solute present, the solution can be made to hold more of the solute than normally is held in equilibrium; the solution is *supersaturated*. Agitation of the solution or the addition of a *seed cyrstal* of solute may start rapid crystallization. After crystallization occurs, a saturated solution will remain. In general, when a solution that is nearly saturated with a solid solute is cooled, a temperature will be reached at which the solution is saturated. On further cooling, the excess solute will crystallize, appearing as particles separated from solution.

While salts are usually more soluble in water at higher temperature, the way the solubility varies with the temperature need not be constant. Thus the change in solubility between 20° and 30° may be different than the change in solubility between 30° and 40°. To fully characterize the solubility properties of a salt, the chemist must experimentally determine the solubility over a wide range of temperatures.

In this experiment you will determine the solubility of $KClO_3$, potassium chlorate, for several temperatures. You can then construct a solubility curve which is a plot of the solubility (in g of solute/100 g solvent) versus the temperature. The plot, or solubility curve, can be used to accurately predict the solubility for any temperature in the range of the experiment.

Experimental Procedure

1. Assemble the apparatus as shown in Figure 36-I. The thermometer is inserted in one hole in the stopper and is adjusted so that the bulb is about 2 cm from the bottom of the test tube. CAUTION: BE VERY CAREFUL NOT TO BREAK THE THERMOMETER DURING THE PROCESS, AS SUCH BREAK-

213

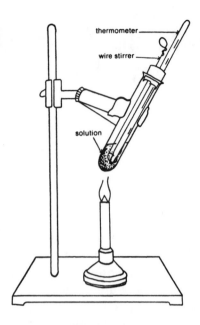
thermometer

wire stirrer

solution

Figure 36-I.

AGE USUALLY IS ACCOMPANIED BY CUT HANDS. To prepare the wire stirrer, bend a loop in one end slightly smaller than the inside diameter of the test tube and pass the other end through the second hole in the rubber stopper. The wire loop should be around the thermometer bulb.

2. Into the clean dry test tube weigh between 4.5-5.0 g of $KClO_3$. Determine the weight of the salt by difference when the apparatus (test tube, thermometer, stopper, and loop) is weighed empty and weighed when it contains the salt. Record all weighings.

3. Add approximately 10 g (10 ml) of water to the apparatus. Determine the exact weight by weighing. Record the weight. Heat the test tube and its contents, stirring continuously with the wire loop. Continue stirring and heating until all of the salt dissolves. Remove the heat and continue to stir. As the solution cools, watch carefully for the appearance of small crystals. When the crystals first start to appear, note the temperature and record it.

4. Cool the solution to room temperature and add a second portion of water of about 2 g (2 ml). Determine the exact weight of water added by weighing the apparatus with its contents. Record the weight. Repeat the heating and cooling cycle, noting and recording the temperature at which crystallization starts.

5. Repeat the procedure of adding water, heating, cooling and noting the temperature at which crystallization starts for a total of six measurements. For the third measurement, add about 8 more g of water; for the fourth measurement, 5 more g of water; for the fifth, 5 g; for the sixth (and last), 10 g.

6. Calculate the concentration (in g salt/100 solvent) of the saturated solution at each temperature.

7. Plot your data on the graph provided. Plot temperature in degrees C on the X-axis and concentration (g salt/100 g water) on the Y-axis.

214

Name_____   Lab Section_____

Station_____   Date_____

The Solubility of a Salt in Water

Weight of apparatus (clean and empty)   _____

Weight of apparatus plus salt   _____

Weight of salt   _____

First water addition

    Weight of apparatus, salt, and water   _____

    Weight of first water added   _____

    Saturation temperature   _____°C

Second water addition

    Weight of apparatus, salt, and water   _____

    Weight of water after second addition   _____

    Saturation temperature   _____°C

Third water addition

    Weight of apparatus, salt, and water   _____

    Weight of water after third addition   _____

    Saturation temperature   _____°C

Fourth water addition

    Weight of apparatus, salt, and water   _____

    Weight of water after fourth addition   _____

    Saturation temperature   _____°C

Fifth water addition

    Weight of apparatus, salt, and water   _____

    Weight of water after fifth addition   _____

    Saturation temperature   _____°C

Sixth water addition

Weight of apparatus, salt, and water   _____

Weight of water after sixth addition   _____

Saturation temperature   _____°C

Determination	Total Weight of water	Solution conc. (g salt/100g H_2O)	Saturation Temperature
1st			
2nd			
3rd			
4th			
5th			
6th			

For the first determination above, show how you calculated the solubility in units of (g salt/100 g water).

From the graph that you construct, what is the solubility of $KClO_3$ at 25°C.

_____ g salt/100 g H_2O

List as many sources of error as you can think of for your experiment, indicating how this would affect your results.

What would happen if you were to place a seed crystal of the solute in an unsaturated solution. —

     a saturated solution —   a supersaturated solution —

Name_____ Section_____

g salt/100 g H$_2$O

Temperature (°C)

Experiment 37
Anion Tests — Known Solutions

Object

In this experiment you will become familiar with the qualitative tests for anions. You will learn how to perform the tests by working with solutions of known composition.

Background

Qualitative analysis of a solution will tell which substances are in the solution; quantitative analysis would tell how much is present in the solution. In today's experiment you will perform qualitative tests on solutions containing known anions (negative ions). You should become familiar with all the tests this week, because in the next experiment you will be required to perform the tests on a solution of unknown composition.

The qualitative test for each anion is a chemical treatment of the solution which will enable you to tell whether that ion is present or absent in the solution. The test for nitrate is the "brown ring test". If the brown ring forms, the solution does contain nitrate ion. If the brown ring does not form, then the solution does not contain the nitrate ion. You must know how to perform the brown ring test, however, to be sure that the absence of the brown ring means no nitrate, rather than indicating an incorrect procedure on your part.

The anions which you will test and a quick description of the test are listed below. The detailed test will be described later.

Sulfate — formation of $BaSO_4$ precipitate in acidic media

Phosphate — formation of ammonium phosphomolybdate precipitate in acid media

Carbonate — add acid to produce carbon dioxide which causes limewater to become cloudy

Chloride — formation of silver chloride precipitate which will dissolve in ammonium hydroxide and reprecipitate when nitric acid is added.

Bromide and Iodide — oxidize iodide to iodine which will give purple color in carbon tetrachloride and then oxidize bromide to bromine which will give red-brown color to the carbon tetrachloride.

Nitrate — brown ring test

Procedure

Perform each of the tests outlined below using 1 ml portions of the sodium salt of the anion being tested. Then repeat the test using several solutions mixed together. Carefully record all observations and then perform any experiments required to answer questions about the test. You should record data and turn in a report.

Sulfate — To a one ml sample of Na_2SO_4 add dilute HCl dropwise until the solution is acidic toward litmus. Add $BaCl_2$ to this solution, formation of $BaSO_4$ as a white precipitate confirms sulfate in the original solution.

To see the importance of adding HCl in the sulfate test, place one ml of Na_2CO_3 in a test tube and add ten drops of $BaCl_2$. Now add HCl dropwise until the precipitate dissolves completely. Repeat this procedure by placing one ml of Na_3PO_4 in another test tube and adding ten drops of $BaCl_2$. Add HCl dropwise until the precipitate dissolves.

Phosphate — Mix ten drops of ammonium molybdate, $(NH_4)_2MoO_4$, with ten drops of dilute HNO_3. If the solution is not clear, see your instructor. Pour one ml of Na_3PO_4 into the molybdate solution. A yellow precipitate of $(NH_4)_3PO_4 \cdot 12MoO_3$ should appear. If the precipitate doesn't appear, place the test tube in a beaker of warm water (50°C) for several minutes. Repeat this test using Na_2SO_4 and Na_2CO_3 as the solutions to be poured into the molybdate solution.

Carbonate — Fit a large test tube with a bent delivery tube and a one-hole rubber stopper as shown in the diagram. Place one ml of Na_2CO_3 in the large test tube. Half fill a medium sized

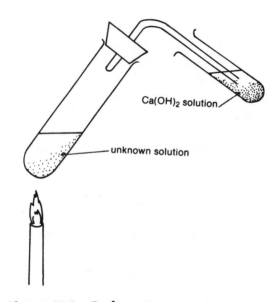

Figure 37-1. Carbonate set up

test tube with clear limewater solution, $Ca(OH)_2$. When the apparatus is ready, quickly add one ml of dilute HCl to the carbonate solution and replace the stopper. Place the tip of the delivery tube in the limewater solution and let the carbon dioxide gas bubble through the limewater. If necessary, you may gently heat the large test tube to expel the gas. Do not permit any of the liquid to boil over. A white precipitate of $CaCO_3$ in the limewater is confirmation of carbonate in the original solution.

Chloride — Place one ml of NaCl in a test tube and add dilute HNO_3 until the solution is acidic to litmus. Add a few drops of $AgNO_3$ and centrifuge the solution. Be sure to counterbalance the solution with a test tube containing an equal volume of water in the centrifuge. Perhaps you may need to mark your test tube with a pencil to avoid confusion.

220

Add one drop of $AgNO_3$ to be sure that precipitation is complete. If the $AgNO_3$ causes cloudiness, repeat the process of centrifuging and adding $AgNO_3$ until no more cloudiness is caused by the addition of $AgNO_3$. Discard the filtrate (the clear solution) by carefully pouring it from the test tube so as to retain the precipitate (the solid). Wash the precipitate with a few ml of distilled water. Pour the distilled water into the test tube and use a small stirring rod to suspend the solid in the liquid; centrifuge and discard the filtrate. Now add three ml of distilled water, 4 drops of dilute NH_4OH and 10 drops of $AgNO_3$. AgCl should dissolve into this solution but AgBr and AgI should not, due to the formation of $Ag(NH_3)_2^+$ from the solid AgCl but not from the very insoluble AgBr and AgI. Confirmation of chloride can be done by treating the clear solution with dilute HNO_3 until it is acidic to litmus paper. AgCl should reprecipitate under these conditions.

Repeat the procedure above using one ml of NaBr. The behavior should be similar to chloride, except that AgBr will not dissolve in the ammonium hydroxide solution. Repeat the procedure using one ml of NaI and similar results should be obtained. Now mix 6 drops of NaCl, 6 drops of NaBr, and 6 drops of NaI and repeat the procedure. Can you detect the AgCl which dissolves into the ammonium hydroxide by the change in the amount of precipitate? Most people will find it necessary to confirm chloride in the presence of bromide or iodide, by the reappearance of AgCl when the ammonium hydroxide solution is acidified with nitric acid. Test 1 ml of Na_2CO_3 with a few drops of $AgNO_3$. Add HNO_3 until the solution is acidic to litmus.

Bromide and Iodide — Place ten drops of NaBr and ten drops of NaI in a test tube. Add 5 drops of carbon tetrachloride, CCl_4. Which liquid is the CCl_4? Add one drop of sodium hypochlorite, NaOCl, and mix well. The iodide should be oxidized to iodine, I_2, under these conditions. The I_2 will dissolve into the CCl_4 layer to give a purple color. Add NaOCl dropwise with vigorous shaking until the purple color disappears. The iodine has been oxidized to iodate, IO_3^-, which is colorless. Add dilute sulfuric acid dropwise until the solution is acidic to litmus paper. Under these conditions the bromide should be oxidized by hypochlorite to bromine, Br_2, which will dissolve into the CCl_4 layer to give a reddish-brown color. Repeat the test using a mixture of NaCl, NaBr and NaI to see if chloride interferes in this test. Repeat the test using a mixture of Na_2CO_3, NaBr and NaI to see how the test is affected by carbonate.

Nitrate — Place about 1 ml of $NaNO_3$ in a test tube and acidify by adding dilute H_2SO_4 dropwise. Litmus paper will turn red when the solution is acidic. Add 1 ml of fresh $FeSO_4$ solution. (Prepare this solution yourself to be sure that it is fresh. Mix solid $FeSO_4$ and distilled water.) Tilt the test tube at a 45° angle and carefully pour about 1 ml of concentrated H_2SO_4 down the side of the test tube. Avoid any undue mixing. Use great care when handling concentrated H_2SO_4. A thin brown ring should appear at the interface of the two layers. Bromide and iodide may interfere in this test. If they are present, free Br_2 and I_2 may be formed at the interface. Repeat the test using a solution containing 1 ml $NaNO_3$ and 1 ml of NaI. Record your observations. To avoid interference by bromide and iodide in the brown ring test, add 2 ml of saturated $AgC_2H_3O_2$ to 1 ml of the test solution. This causes bromide and iodide to be precipitated as AgBr and AgI. Add a few drops of dilute HCl to precipitate any excess silver ions. Centrifuge the mixture and perform the nitrate test on the solution as before. Practice this procedure until you can obtain a suitable nitrate test on a solution containing $NaNO_3$, NaBr and NaI.

Describe the results of performing the nitrate test on a solution made by mixing 1 ml of $NaNO_3$ and 1 ml of Na_2CO_3.

Name_____   Section_____

Station_____   Date_____

Data Sheet

Data

Sulfate Test.

(a) Describe results of mixing Na_2SO_4, HCl and $BaCl_2$.

(b) Describe results of mixing Na_2CO_3 and $BaCl_2$. How is this precipitate affected by HCl?

(c) Describe the results of mixing Na_3PO_4 and $BaCl_2$. How is this precipitate affected by HCl?

Phosphate Test.

(a) Describe the precipitate formed by mixing the molybdate and phosphate solutions.

(b) Did any other ions cause a precipitate with the molybdate?

223

Carbonate Test

(a) Describe the appearance of the limewater solution before and after the carbonate test.

(b) What happened to the carbonate solution when you added HCl?

(c) Do any other anions show this behavior?

Chloride Test

(a) Describe the precipitate which you obtain when NaCl, HNO_3 and $AgNO_3$ are mixed.

(b) Describe the effect of adding NH_4OH and $AgNO_3$ to the precipitate.

(c) Describe the effect of adding HNO_3 to the clear solution.

(d) Describe the precipitate you obtain when NaBr, HNO_3 and $AgNO_3$ are mixed. How did it behave toward NH_4OH?

Data Sheet

Name_____   Section_____

(e) Describe the precipitate you obtain when NaI, HNO_3 and $AgNO_3$ are mixed. How did it behave toward NH_4OH?

(f) Describe the results you obtained with the $NaCl$, $NaBr$, NaI mixture.

(g) From the results of mixing Na_2CO_3 and $AgNO_3$ describe the solubility of Ag_2CO_3 in water.

(h) How can you distinguisn between $AgCl$ and Ag_2CO_3?

Bromide and Iodide Test

(a) Describe the positive results of an iodide test.

(b) How much hypochlorite is required to oxidize iodine to iodate? Is it possible to miss iodide in an unknown if you use too much hypochlorite?

(c) Describe the positive bromide test.

(d) What is the difference in conditions required to oxidize bromide to bromine and iodide to iodine using hypochlorite as the oxidizing agent.

225

Nitrate

(a) What happens to $FeSO_4$ solutions which have been in contact with air a long time?

(b) Describe the positive nitrate test you obtained with $NaNO_3$.

(c) Describe the test you performed with $NaNO_3$ and NaI.

(d) Describe the procedure you found necessary to test for nitrate in the presence of bromide or iodide.

(e) How does carbonate interfere in the nitrate test? How could this interference be avoided?

Experiment 38
Anion Unknown Solution

Object

In this experiment you will perform qualitative tests for anions in solution of unknown composition.

Background

You should now be familiar with the qualitative tests for the anions you have studied: sulfate, phosphate, carbonate, chloride, bromide, iodide, and nitrate. You should know how to perform the test on a solution which contains only one ion. You should also know which ions interfere in each test and the procedures to avoid or eliminate any problem in performing any test in the presence of other ions. You will now test an unknown solution to determine which anions it contain.

Procedure

Turn in a clean dry test tube in return for a numbered test tube filled with an unknown solution. Record the number of the test tube. Perform anion tests on separate 1 ml portions of the unknown. Record all your observations and turn in your report at the end of the lab period.

Name_____   Section_____

Station_____   Date_____

Anion Unknown
Report

Anions confirmed in tests are circled in the list:

SO_4^{2-} PO_4^{3-} CO_3^{2-} Cl^- Br^- I^- NO_3^-

Unknown Number_____.

Observations in each test:

1. <u>Sulfate Test</u>

     Sulfate is present/absent (circle one).

2. <u>Phosphate Test</u>

     Phosphate is present/absent (circle one).

3. <u>Carbonate Test</u>

     Carbonate is present/absent (circle one)

4. <u>Chloride Test</u>

Chloride is present/absent (circle one)

5. <u>Bromide Test</u>

Bromide is present/absent (circle one)

6. <u>Iodide Test</u>

Iodide is present/absent (circle one)

7. <u>Nitrate Test</u>

Nitrate is present/absent (circle one)

Experiment 39
Cations — Group I

Object

In this experiment you will perform qualitative analysis on an unknown solution which contains cations in Group I. The cations in Group I are silver, lead and mercurous, Ag^+, Pb^{2+} and Hg_2^{2+}, respectively; all form insoluble chlorides.

Background

Many chemistry students have studied qualitative analysis of cations using a scheme which separates the cations into Groups according to some common chemical property. Classically, Group I contains cations which will precipitate upon the addition of HCl. Most chlorides are soluble, however, Ag^+ and Hg_2^{2+} (mercurous which occurs as two Hg^+ ions bonded into a diatomic ion, Hg_2^{2+}) chlorides are quite insoluble. The chloride of lead is soluble in hot water but insoluble at room temperature.

After the group of cations are precipitated, the qualitative tests must be continued to confirm which of the cations are presnt. These tests indicate the presence or absence of each ion and rely on specific color or solubility of substances formed in the procedure.

The tests are more complex than those employed in the anion tests, in a sense, because prior treatment of the sample is important to eliminate interfering ions. To display the sequence of tests to be performed on the Group I of cations, we find it useful to employ a Flow Diagram. This is an abbreviated description of the reagents to be used and the results which will confirm each ion. It does not indicate the amount of reagent to use; the detailed directions for the tests will be listed separately. The Flow Diagram for Group I is listed on page 232.

The Diagram shows that the unknown is treated with HCl (step 1) which causes a precipitate to form. The right branch represents the solution or filtrate and the left branch represents the precipitate. If you are testing an unknown which contains only ions from Group I, you may discard the filtrate since all the cations of interest are in the precipitate. Substances which precipitate are indicated by a downward arrow, ↓. In step 2, the precipitate is treated with hot water which causes $PbCl_2$ to dissolve. This solution is treated with K_2CrO_4 (step 3) and the formation of $PbCrO_4$ as a yellow precipitate confirms the presence of lead in the unknown. Confirmation tests are underlined. In this case the solution will be yellow due to chromate ions; lead is confirmed only by the formation of the yellow precipitate, $PbCrO_4$.

The precipitate from step 2 is treated with NH_4OH. Formation of a grey precipitate here, which is a mixture of black Hg and white $Hg(NH_2)Cl$, confirms mercurous in the unknown. The solution is treated with HNO_3 and reprecipitation of AgCl is confirmation of silver in the unknown.

231

GROUP I – FLOW DIAGRAM

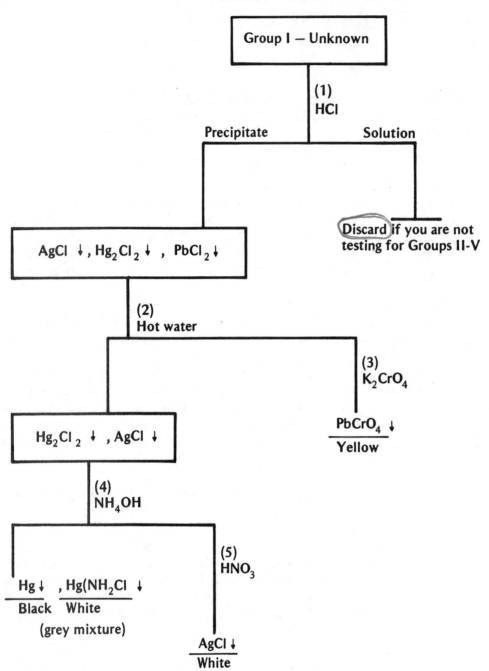

Group I – Unknown

(1)
HCl

Precipitate | Solution

Discard if you are not testing for Groups II-V

$AgCl \downarrow$, $Hg_2Cl_2 \downarrow$, $PbCl_2 \downarrow$

(2)
Hot water

(3)
K_2CrO_4

$PbCrO_4 \downarrow$
Yellow

$Hg_2Cl_2 \downarrow$, $AgCl \downarrow$

(4)
NH_4OH

(5)
HNO_3

$Hg \downarrow$, $Hg(NH_2Cl \downarrow$
Black   White
(grey mixture)

$AgCl \downarrow$
White

Procedure

Perform the tests on Group I using <u>one ml</u> of $AgNO_3$; then perform the tests on one ml of $Pb(NO_3)_2$ and then on one ml of $Hg_2(NO_3)_2$. Then repeat using mixtures of these substances. When you are sure that you have mastered each test, trade a clean dry test tube for a numbered test tube filled with an unknown solution. Use 3 ml of the unknown for the tests. Record your observations and turn in your report before you leave the laboratory.

1. To ~~3 ml~~ [1 ml] of the solution to be tested, add dilute HCl dropwise until precipitation is complete. Centrifuge and save the precipitate. You may discard the filtrate if you are not testing for Groups II-V. If you are testing for these groups, save the filtrate since it contains these cations.

2. Add 5 ml of hot water to the precipitate, stir with a glass rod which has been warmed in hot water. Centrifuge quickly and separate the solution and precipitate. [Pour into test tube]

3. Add a few drops of K_2CrO_4 to the filtrate. A yellow precipitate of $PbCrO_4$ confirms lead, Pb^{2+}, in the unknown. [seperate test tube]

4. Add 1 ml of dilute NH_4OH [$NH_3 + H_2O$] and 2 ml distilled water to the precipitate from step 2. Mix the contents of the test tube well and centrifuge. A grey precipitate confirms mercurous. The precipitate may appear black due to finely divided Hg or nearly white due to $Hg(NH_2)Cl$.

5. Pour the filtrate from step 4 into a clean test tube and add dilute HNO_3. Formation of a white precipitate, AgCl, confirms silver in the unknown.

Unknown

Pb   went to bottom

Ag   some float on top

Pb + Ag

H_2O

Experiment 40
Cations — Group II

Object

In this experiment you will perform qualitative analysis on an unknown solution which contains the cations in Group II. These cations are ones which form sulfides which are insoluble in acidic solution, mercuric, lead, bismuth, and copper, Hg^{2+}, Pb^{2+}, Bi^{3+} and Cu^{2+}.

Background

Many qualitative analysis schemes have relied on hydrogen sulfide, H_2S, as the precipitating agent. This noxious and toxic gas is difficult to handle. We will employ thioacetamide, $CH_3C(S)NH_2$, to produce H_2S in the solution by a slow hydrolysis reaction shown in the equation:

$$CH_3C(S)NH_2 + H_2O \longrightarrow CH_3C(O)NH_2 + H_2S$$

Thioacetamide is a clean and convenient source of hydrogen sulfide.

The cations in Group II, Hg^{2+}, Pb^{2+}, Bi^{3+} and Cu^{2+} form sulfides which are quite insoluble, and will therefore form precipitates even with very low sulfide ion concentrations. The concentration of sulfide ion in a solution of hydrogen sulfide varies with the concentration of hydrogen ion. This can be explained by writing the equilibrium:

$$H_2S \rightleftharpoons 2H^+ + S^{2-}$$

By applying Le Chateler's Principle to this system, we can see that increasing the concentration of H^+ in a solution of constant H_2S concentration will decrease the concentration of sulfide ion in the solution. We will adjust the concentration of H^+ in the solution to 0.3 M and saturate the solution with H_2S (about 0.1 M solution). This will give a sulfide concentration of 1.4×10^{-20} M. Virtually all of the ions in Group II will be precipitated under these conditions.

The Flow Chart for Group II is found on page 238. The solution is treated with HCl and H_2S to precipitate the sulfides. The filtrate may be discarded if you are testing an unknown which contains only ions in Group II. The precipitate is treated with dilute nitric acid and the solution is heated (step 7). All the sulfides, except HgS, should dissolve under these conditions. The HgS precipitate which remains is treated with aqua regia (a mixture of HNO_3 and HCl) and will dissolve due to the formation of $HgCl_4^{2-}$. The solution is separated from the residue and boiled to remove nitrogen oxides which are oxidizing agents. The solution is treated with stannous chloride, $SnCl_2$, which is a good reducing agent. The $SnCl_2$ will cause the $HgCl_4^{2-}$ to be reduced to black Hg or white Hg_2Cl_2. Formation of this grey, white or black precipitate confirms mercuric ion in the original unknown.

The solution from step 7 is treated with H_2SO_4 and heated strongly to remove volatile HCl and HNO_3. This mixture is treated with water but $PbSO_4$ will remain as a precipitate (step 9). The $PbSO_4$ is dissolved in ammonium acetate, $NH_4C_2H_3O_2$, in step 10. Confirmation of lead in the unknown is done by adding K_2CrO_4 and obtaining the yellow $PbCrO_4$ as a precipitate. The solution from step 9 is

237

treated with NH_4OH. Confirmation of copper in the unknown is the production of the deep blue color due to $Cu(NH_3)_4^{2+}$ in the solution. The precipitate from step 11 is white $Bi(OH)_3$. This precipitate will yield metallic Bi upon treatment with Sn^{2+}. Confirmation of bismuth in the unknown is the production of the black precipitate, Bi, upon the addition of sodium stannite.

GROUP II — FLOW DIAGRAM

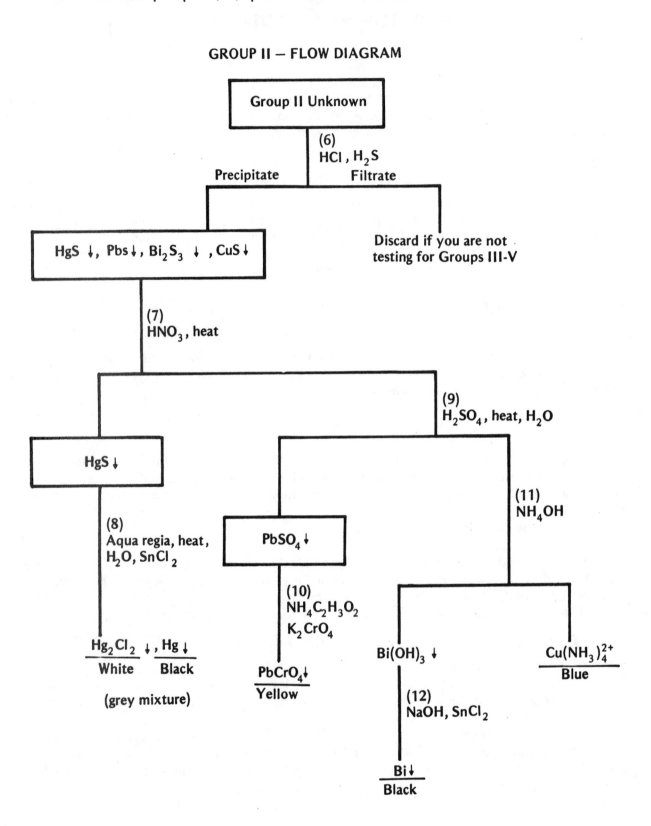

Procedure

This group of cations is more difficult to test than the previous one. You will *not* have time to test all known solutions before testing the unknown solution. It is suggested that you test a mixture of $Hg(NO_3)_2$ and $Pb(NO_3)_2$ first. When you are familiar with these tests, obtain an unknown solution and perform all the tests on it. If you encounter any trouble, see if any tests on known solutions may help you. If you cannot solve the problem by yourself, consult with your teaching assistant for further help.

6. Neutralize the solution to litmus with dilute HCl or dilute NH_4OH. Add one drop of dilute HCl for each ml of unknown solution. The concentration of H^+ should be 0.3 M if you follow this procedure. Use 2 ml of this unknown solution for testing. Add ten drops of thioacetamide solution and heat the test tube in a hot water bath for five minutes. Cool, and centrifuge; you may discard the filtrate if you are not testing for Groups III-V. Wash the precipitate with one drop NH_4Cl, one drop thioacetamide and ten drops distilled water. Discard the solution and retain the precipitate.

7. Add five drops of dilute HNO_3 and five drops of distilled water to the precipitate. Stir up the precipitate and place in a hot water bath for five minutes. A black precipitate at this point indicates HgS, but mercuric ion should be confirmed by step 8. The other ions should be in solution.

8. Add one drop of concentrated HNO_3 and three drops of concentrated HCl to the precipitate from step 7. This HNO_3/HCl mixture is called aqua regia, it is a very powerful oxidizing and complexing agent. Stir the precipitate and place in boiling water for five minutes. Cool the solution. Add ten drops of water and three drops of stannous chloride, $SnCl_2$. Formation of white Hg_2Cl_2 or black Hg or a grey mixture of Hg_2Cl_2 and Hg confirms mercuric ion, Hg^{2+} in the unknown.

9. Put four drops of concentrated H_2SO_4 into the filtrate from step 7. Transfer to a clean crucible and heat carefully with a bunsen burner. As the solution is concentrated, HCl and HNO_3 will be expelled. When dense white fumes of SO_3 begin coming off of the sample, stop heating the sample and allow it to cool. Add ten drops of water and transfer the solution to a clean test tube. You may find it necessary to scratch the test tube with a glass rod to cause the precipitation of white $PbSO_4$. Save the filtrate and wash the precipitate with five drops of water.

10. Dissolve the precipitate in four drops of ammonium acetate, $NH_4C_2H_3O_2$. Warm the solution if necessary. Add two drops of potassium chromate, K_2CrO_4. The formation of the yellow precipitate, $PbCrO_4$, confirms lead, Pb_2^+, in the unknown.

11. Neutralize the solution from step 9 with NH_4OH until it is distinctly basic toward litmus. Formation of the deep-blue color of $Cu(NH_3)_4^{2+}$ in the solution is confirmation of copper, Cu^{2+} in the unknown. Formation of a white precipitate is an indication of bismuth in the unknown. Bismuth should be confirmed by step 12.

12. Wash the precipitate from step 11 with ten drops of water; discard the water. Prepare a fresh solution of sodium stannite, $Na_2Sn(OH)_4$ by mixing two drops of $SnCl_2$ with 6 M NaOH. Add just enough 6 M NaOH to cause the solution to become clear after the initial precipitation of $Sn(OH)_2$. Pour the $Na_2Sn(OH)_4$ solution on the $Bi(OH)_3$ precipitate. Formation of finely divided bismuth, which appears black, confirms bismuth, Bi^{3+} in the unknown.

Qualitative Analysis

Flame Test

Na - Orange / Yellow

K - Purple

Ca - Red

Ba - Yellow

Cu^{2+} - Blue   + NH_3 - Deep Blue

Fe^{3+} - Yellow/Orange

  + SCN^- - Redish brown

  + NH_3 - Red precipitate formed

Ca^{2+} - White precipitate at bottom kinda chalky

Zn^{2+} - Very white not too solid of precipitate white soln.

Name_____   Section_____

Station_____   Date_____

Group II Unknown

Report

Observations on Known Solutions.

$Pb(NO_3)_2 - Hg(NO_3)_2$ mixture

Step 6.

Step 7.

Step 8.

Step 9.

Step 10.

Step 11.

Step 12.

Other tests or observations.

Observations on Unknown Solution Number_____ .

Step 6.

Step 7.

Step 8.

Step 9.

Step 10.

Step 11.

Step 12.

Hg^{2+} present/absent (circle one). Explain why you conclude this.

Pb^{2+} present/absent (circle one). Explain why you conclude this.

Cu^{2+} present/absent (circle one). Explain why you conclude this.

Bi^{3+} present/absent (circle one). Explain why you conclude this.

Experiment 41
Cations — Group III

Object

In this experiment you will perform qualitative analysis on an unkown containing cations in Group III. These cations precipitate from an alkaline sulfide solution and include manganous, zinc, iron and chromium ions, Mn^{2+}, Zn^{2+}, Fe^{2+} (or Fe^{3+}) and Cr^{3+}.

Background

The conditions used to precipitate this group of cations are mildly alkaline. The solution contains both NH_4Cl and NH_4OH and is therefore a buffer solution. The concentration of hydroxide in this mixture is near the value of ionization constant of NH_4OH or $10^{-5}M$. The H^+ concentration will be near $10^{-8}M$. From the dissociation equation of H_2S, it can be seen that high sulfide ion concentrations will be obtained in solutions of low hydrogen ion concentration. When the H^+ concentration is $10^{-8}M$ and the solution is saturated with H_2S gas, a S^{2-} concentration of about $10^{-5}M$ will be obtained. You can see that a much higher sulfide ion concentration is used to precipitate Group III than was used for Group II.

The sulfides of the Group II cations are more insoluble than the sulfides of the Group III cations. Under the conditions employed, chromic ions would be precipitated as $Cr(OH)_3$ and not as the sulfide. Ferric ions, Fe^{3+}, would be precipitated as $Fe(OH)_3$; however, Fe^{3+} is readily reduced to Fe^{2+}, ferrous ions, by sulfide. Ferrous ions would be precipitated as FeS.

The Flow Diagram for Group III is shown on page 244. The unknown solution is treated with NH_4Cl and NH_4Cl and H_2S is added. The precipitate will contain MnS, ZnS, FeS, and $Cr(OH)_3$. The precipitate is dissolved in HCl. When the solution is made alkaline and treated with hydrogen peroxide, H_2O_2, the following processes will occur:

(a) Mn^{2+} is oxidized to Mn^{4+} and precipitated as MnO_2
    (MnO_2 can be considered as dehydrated $Mn(OH)_4$)

(b) Fe^{2+} is oxidized to Fe^{3+} and precipitated as $Fe(OH)_3$

(c) Zn^{2+} forms amphoteric $Zn(OH)_2$ which dissolves in excess base by forming zincate ion, $Zn(OH)_4^{2-}$.

(d) Cr^{3+} is oxidized to yellow CrO_4^{2-}, chromate, ion.

The solution which contains CrO_4^{2-} and $Zn(OH)_4^{2-}$ is neutralized with HCl and made alkaline with NH_4OH. The chromate is precipitated by the addition of barium acetate, $Ba(C_2H_3O_2)_2$. Formation of yellow barium chromate, $BaCrO_4$, precipitate is confirmation of chromium in the original unknown. The solution is treated with H_2S and the formation of white zinc sulfide, ZnS, as a precipitate is confirmation of zinc in the unknown.

The precipitate containing MnO_2 and $Fe(OH)_3$ is dissolved in HCl. Half of the solution is treated with nitric acid and sodium bismuthate, $NaBiO_3$, which is such a powerful oxidizing agent that it will oxidize Mn(IV) to Mn(VII). Formation of the purple color of permanganate, MnO_4^-, is confirmation of manganese in the unknown. The other half of the solution is treated with potassium thiocyanate, KCNS. Formation of the red-brown color characteristic of ferric thiocyanate, $FeSCH^{2+}$, in the solution is confirmation of iron in the unknown.

GROUP III — FLOW DIAGRAM

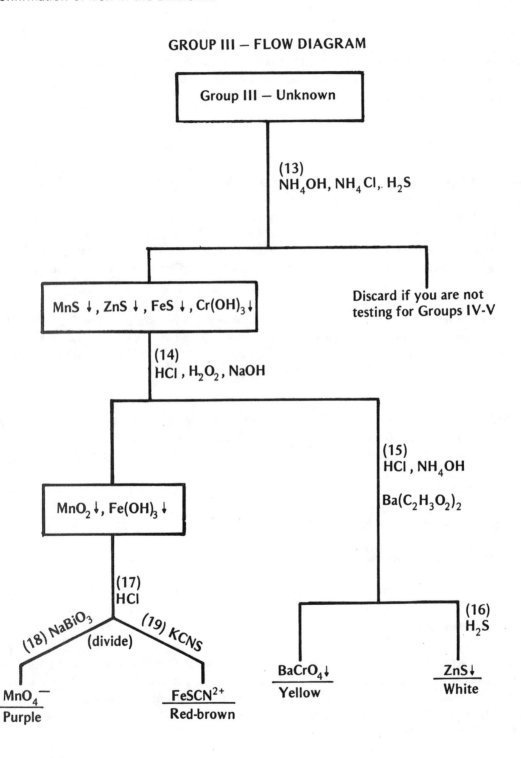

Procedure

The tests for the Group III cations are not as complex as those of Group II. You should have time in this experiment to test a known mixture which contains all of the ions before you attempt to test an unknown solution.

(13) Treat 2 ml of the unknown solution with four drops of dilute HCl. Add dilute NH_4OH dropwise until the solution is distinctly alkaline toward litmus. Add 15 drops of thioacetamide and warm the test tube in a boiling water bath for five minutes. Cool the solution and centrifuge. You may discard the filtrate if you are not testing for Groups IV and V.

(14) The precipitate from step 13 contains MnS, ZnS, FeS, $Cr(OH)_3$ and possibly some $Fe(OH)_3$. Wash this precipitate with distilled water; discard the water. Dissolve the precipitate in 5 drops of concentrated HCl. Add several more drops of HCl if necessary. If the precipitate will not dissolve completely, centrifuge and discard the precipitate. Perform the tests on the clear filtrate. Add 6 M NaOH dropwise to the filtrate until it is alkaline to litmus and then add 5 drops more NaOH. Add 6 drops of 5% H_2O_2 and place in boiling water bath for five minutes. Centrifuge the mixture. Save the precipitate for step 17.

(15) Add concentrated HCl dropwise to the filtrate from step 14 until the solution is acidic to litmus. Add concentrated NH_4OH dropwise until the solution is alkaline to litmus, and then add 5 drops excess. Add barium acetate, $Ba(C_2H_3O_2)_2$, dropwise to the solution until precipitation of $BaCrO_4$ is complete. Formation of the yellow $BaCrO_4$ precipitate is confirmation of chromium in the unknown.

(16) Add 5 drops of thioacetamide to the solution from step 15 and place in a boiling water bath for 5 minutes. Formation of a white precipitate of zinc sulfide, ZnS, is confirmation of zinc in the unknown.

(17) Wash the precipitate from step 14 (MnO_2 and $Fe(OH)_3$) with 5 drops of distilled water; centrifuge and discard the water. Dissolve the precipitate in a minimum volume of concentrated HCl. Add 2 ml of distilled water. Mix the solution and divide it into two equal parts.

(18) To half of the solution from step 17, add a small portion (about the size of the head of a kitchen match) of solid sodium bismuthate, $NaBiO_3$. Stir the mixture well and then centrifuge. Appearance of the purple color of permanganate, MnO_4^-, in the clear solution is confirmation of manganese in the unknown.

(19) To half of the solution from step 17, add two drops of potassium thiocyanate, KCNS. Formation of reddish-brown (nearly blood red) ferric thiocyanate complex, $FeSCN^{2+}$, is confirmation of iron in the unknown.

Name_____   Section_____

Station_____   Date_____

Group III

Unknown Report

Observations on known solution.

Step 13.

Step 14.

Step 15.

Step 16.

Step 17.

Step 18.

Step 19.

Observations on unknown solution number_____.

Step 13.

Step 14.

Step 15.

Step 16.

Step 17.

Step 18.

Step 19.

Cr^{3+} present/absent (circle one). Explain why you conclude this.

Zn^{2+} present/absent (circle one). Explain why you conclude this.

Mn^{2+} present/absent (circle one). Explain why you conclude this.

Fe^{2+} or Fe^{3+} present/absent (circle one). Explain why you conclude this.

Experiment 42
Cations — Groups IV and V

Object

In this experiment you will perform qualitative analysis on an unknown containing cations in Groups IV and V. The cations in Group IV are calcium and barium, Ca^{2+} and Ba^{2+}, which form insoluble carbonates. The cations in Group V are sodium, Na^+, and ammonium, NH_4^+, which form no insoluble substances.

Background

The cations studied in this experiment are calcium and barium, Ca^{2+} and Ba^{2+}, in Group IV and ammonium and sodium, NH_4^+ and Na^+ in Group V. The ammonium test is performed first on a separate portion of the unknown. The unknown is treated with NaOH and tested for NH_3 gas by moist litmus paper. The unknown solution is buffered with NH_4OH and a small amount of NH_4Cl and treated with ammonium carbonate, $(NH_4)_2CO_3$. The Group IV cations will precipitate and the solution is tested for sodium ion using a flame test.

The $BaCO_3$ and $CaCO_3$ are dissolved in acetic acid and the solution is buffered to a pH of 5. Barium is precipitated from the solution as $BaCrO_4$ by adding K_2CrO_4. Barium is confirmed by a flame test and by the formation of barium sulfate, $BaSO_4$. Calcium is confirmed by a flame test and by the formation of calcium oxalate, CaC_2O_4, as a precipitate.

The Flow Chart for these cations is found on page 250.

Procedure

Perform these tests on a solution known to contain all of the ions. Pay particular attention to the flame tests. For each flame test, work with the solution being tested, distilled water, a solution known to contain the ion, distilled water, and a solution which does not contain the ion.

(20) Ammonium test is performed on a small portion of the original unknown. Place five drops of the unknown in a clean, dry test tube. Moisten a piece of red litmus paper with distilled water. Pour 1 ml of 6 M NaOH into the unknown and place the strip of moist litmus paper in the neck of the test tube. Ammonium ion in the unknown will produce ammonia gas in this test, which will cause the moist litmus paper to turn blue.

(21) Place 2 ml of the unknown in another test tube and add 4 drops of 1 M NH_4Cl and 4 drops of dilute NH_4OH. Add 6 drops of 2.5 M ammonium carbonate, $(NH_4)_2CO_3$. Place the test tube in a boiling water bath for 10 minutes. Cool and centrifuge. Save the filtrate for the sodium flame test. Test the precipitate for calcium and barium.

(22) The solution from step 21 will contain ammonium ion and may contain sodium ion. Since sodium salts don't have characteristic colors and since most are soluble, we will use a flame test for sodium. Obtain a nichrome wire loop from the stockroom. Soak it in distilled water, and heat it in a clean,

GROUPS IV & V – FLOW DIAGRAM

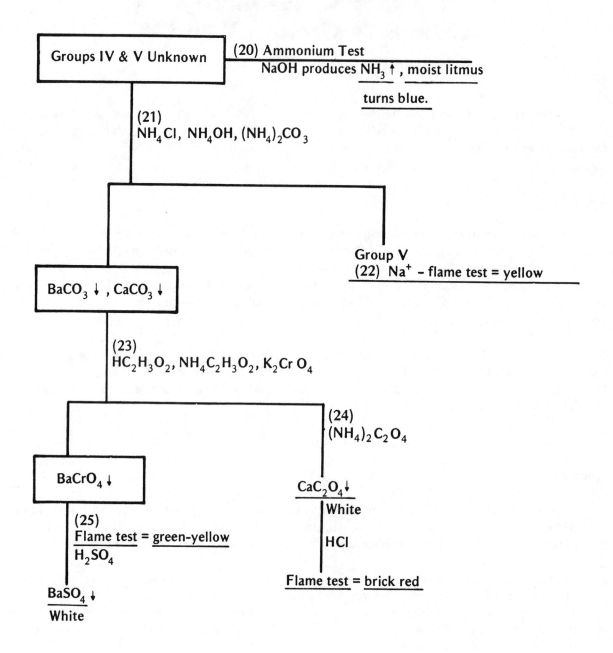

Groups IV & V Unknown

(20) Ammonium Test
NaOH produces $NH_3 \uparrow$, moist litmus turns blue.

(21)
NH_4Cl, NH_4OH, $(NH_4)_2CO_3$

Group V
(22) Na^+ – flame test = yellow

$BaCO_3 \downarrow$, $CaCO_3 \downarrow$

(23)
$HC_2H_3O_2$, $NH_4C_2H_3O_2$, K_2CrO_4

(24)
$(NH_4)_2C_2O_4$

$BaCrO_4 \downarrow$

$CaC_2O_4 \downarrow$
White

(25)
Flame test = green-yellow
H_2SO_4

HCl

Flame test = brick red

$BaSO_4 \downarrow$
White

blue bunsen burner flame. Dip it in nitric acid and then heat it to red hot temperatures in the flame. Repeat the process until the wire is perfectly clean and does not glow with any characteristic color while being heated. When the wire is clean, dip it in the unknown solution, and place a small drop of the unknown in the flame. Sodium in the unknown will cause a bright yellow glow in the flame. Be sure to practice this test until you can distinguish between a true sodium test and slight sodium contamination. Most solutions will have traces of sodium present (from glass, sweat, tap water, etc.) and you must be able to spot the difference.

(23) Dissolve the $CaCO_3$, $BaCO_3$ precipitate with stirring in 6 M acetic acid. Add 5 drops of ammonium acetate, $NH_4C_2H_3O_2$, and add distilled water to give a total volume of 2 ml. Add K_2CrO_4 dropwise until precipitation is complete and centrifuge. To be sure that all barium is precipitated, add one drop of K_2CrO_4 in excess. Centrifuge and be sure to retain both the precipitate and the filtrate.

(24) Add dilute NH_4OH to the filtrate from step 23 until the solution changes from orange ($Cr_2O_7^{2-}$ in acidic medium) to yellow (CrO_4^{2-} in basic medium). Add 10 drops of ammonium oxalate, $(NH_4)_2C_2O_4$, and warm the solution in a hot water bath for 10 minutes. Formation of white calcium oxalate, CaC_2O_4, as a precipitate can be considered as confirmation of calcium in the unknown. As further confirmation, the precipitate of CaC_2O_4 can be dissolved in dilute HCl and this solution used in a flame test. Calcium will impart a brick-red color to the clean nichrome wire.

(25) Dissolve the precipitate from step 23, $BaCrO_4$, in 6 drops of concentrated HCl. Barium should be confirmed in this solution by a flame test and by precipitation of $BaSO_4$. The HCl solution should cause a faint and brief green-yellow color to appear on the nichrome wire held in the flame. Add ten drops of dilute H_2SO_4 to the solution and the white precipitate, $BaSO_4$, should appear after a short while.

Name_____ Section_____

Station_____ Date_____

Groups IV & V

Unknown Report

Observations on solution containing all ions.

Step 20.

Step 21.

Step 22.

Step 23.

Step 24.

Step 25.

253

Observations on unknown solution number_____ .

Step 20.

Step 21.

Step 22.

Step 23.

Step 24.

Step 25.

NH_4^+ present/absent (circle one). Explain why you conclude this.

Na^+ present/absent (circle one). Explain why you conclude this.

Ca^{2+} present/absent (circle one). Explain why you conclude this.

Ba^{2+} present/absent (circle one). Explain why you conclude this.

Experiment 43
General Cation Unknown

Object

In today's experiment you will perform qualitative analysis on an unknown solution containing any of the fourteen cations you have previously studied in Groups I-V.

Background

In recent experiments you have learned qualitative analysis tests for fourteen different cations divided into five different groups. In the earlier experiments you studied only one group of cations during each laboratory period. You analyzed unknown solutions which contained only cations in the group you were investigating. In today's experiment you will perform the same tests on an unknown solution which may contain any of the cations you have already studied.

Your procedure will involve adding a reagent to the unknown to remove a group of cations. You must be careful to adjust the amounts of reagent to completely remove the cations from the unknown. In the case of Group I cations, for example, enough HCl must be added to remove essentially all of the Ag^+ and Hg_2^{2+} from the unknown solution. The filtrate will be saved for tests on the cations in Groups II-V. If the Ag^+ and Hg_2^{2+} are not removed from the filtrate, they will interfere with later tests. The next step in the procedure would be to precipitate the Group II cations; they must be completely removed so that they don't interfere with later tests.

Steps 1, 6, 13 and 21 are Group Precipitation reactions. Remember that the filtrate from each of these steps must be saved for further tests. In the past experiments you could discard this, but now that solution must be saved for testing of cations in the later groups. It may also be necessary to concentrate this solution by boiling off excess water.

The Flow Diagram for the General Cation Unknown is shown on page 257. The Group Precipitation reactions are shown with the same step numbers as in previous experiments. The treatment is basically the same as you have used previously; however, it may be necessary to modify the treatment slightly. The modifications are described in this experiment.

The unknowns issued with this experiment will contain only three, four or five cations. You will therefore obtain more negative results than you have encountered before. This will save you time if you really understand which steps you can omit when you obtain certain negative results. For example, if your unknown does not give any precipitate when you add HCl, you may omit all tests for Group I and the tests involving Pb^{2+} in Group II. If you have confirmed Pb^{2+} in Group I tests, there is no need to perform any confirmation tests for Pb^{2+} in Group II.

Procedure

Perform the ammonium test on a small portion of the unknown solution. Perform the remaining tests on 3 ml of the unknown solution. Some of the unknown solution should be left; save this in case you have trouble. If you have trouble during the tests, for example if you throw away a filtrate containing some ions, refer to step number 26.

255

The steps listed below refer to the Group Separation Steps as modified to be used on a general unknown solution. Each precipitate should be tested using the same procedure as in earlier experiments.

Record all observations.

(20) Ammonium Test. Place 5 drops of the unknown in a clean dry test tube. Pour 1 ml of 6 M NaOH into the test tube and test the neck of the test tube with moist, red litmus paper. If the paper turns blue from the gases evolved, ammonium ion is indicated.

(1) Group I Precipitation. Place a 3 ml portion of the unknown solution in a test tube. Add dilute HCl dropwise until precipitation is complete. Be sure to centrifuge so that you are observing a clear solution when testing to see that the HCl causes no further precipitate. Centrifuge, test the precipitate for Group I cations by Steps 2-5 on page 233; retain the filtrate for tests on Groups II-V.

(6) Group II Precipitation. Place the filtrate from step 1 in clean crucible and boil it down to a volume of about 1 ml. Cool the solution and return it to a test tube. Add dilute NH_4OH dropwise until solution is neutral to litmus. Add 2 drops of dilute HCl and 5 drops of thioacetamide solution. Place the test tube in a boiling water bath for 5 minutes. Centrifuge and add two drops of thioacetamide solution. If a precipitate forms, add three drops of thioacetamide and heat for another 5 minutes. Be sure that all Group II cations are precipitated. Perform the Group II tests on the precipitate using steps 7-12 on page 239. Save the filtrate for the Group II-V tests. If you don't plan to use this solution within 15 minutes, gently boil it for two minutes to remove excess H_2S.

(13) Group III Precipitation. To the filtrate from step 6, add dilute NH_4OH until the solution is distinctly basic to litmus paper. Add 10 drops of thioacetamide solution and place the test tube in a boiling water bath for five minutes. Centrifuge the mixture. Add HCl to the filtrate until it is acidic to litmus. Pour this solution in a clean crucible and gently boil off water until the volume is about 2 ml. Test the precipitate for Group III cations by steps 14-19 on page 245.

(21) Group IV Precipitation. Pour the 2 ml of solution from step 13 into a clean test tube. Centrifuge and discard the precipitate. Add 4 drops of dilute NH_4OH (first), 4 drops of 1 M NH_4Cl, and 6 drops of $(NH_4)_2CO_3$ (finally). Warm the solution for 10 minutes in a hot water bath. Centrifuge the mixture. Test the precipitate for Group IV cations by steps 23-25 on page 251. Test the solution for sodium ion by the flame test described in step 22 on page 249.

(26) In Case of Trouble. In case you discard a filtrate or a precipitate before you have tested it completely — don't panic. If you understand the tests you are performing, you can easily continue your work using 3 ml of the original unknown solution. You don't need to repeat all of the procedures. You should be able to quickly obtain the solution or precipitate you need. Suppose you need the Group II precipitate. Add HCl to the unknown solution, then add thioacetamide and heat for 5 minutes. Centrifuge the mixture. You may discard the precipitate since it contains Group I and Group II ions. Treat the filtrate as in step 13 to precipitate the Group III cations.

If you need the filtrate from step 13 for testing, add HCl and then NH_4OH to the unknown solution. Add thioacetamide and heat the mixture; the precipitate will contain ions in Group I, II and III. The filtrate will be the same as the filtrate obtained from step 13.

FLOW DIAGRAM FOR GENERAL CATION UNKNOWN

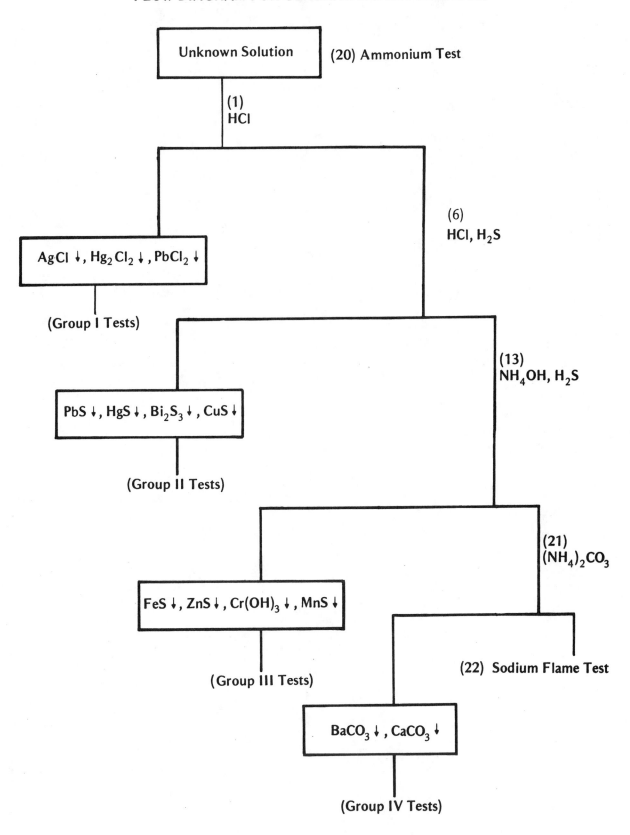

Unknown Solution — (20) Ammonium Test

(1) HCl

$AgCl \downarrow, Hg_2Cl_2 \downarrow, PbCl_2 \downarrow$

(Group I Tests)

(6) HCl, H_2S

$PbS \downarrow, HgS \downarrow, Bi_2S_3 \downarrow, CuS \downarrow$

(Group II Tests)

(13) NH_4OH, H_2S

$FeS \downarrow, ZnS \downarrow, Cr(OH)_3 \downarrow, MnS \downarrow$

(Group III Tests)

(21) $(NH_4)_2CO_3$

(22) Sodium Flame Test

$BaCO_3 \downarrow, CaCO_3 \downarrow$

(Group IV Tests)

257

Name_____ Section_____

Station_____ Date_____

General Unknown

Observations on unknown number_____.

Step 20. NH_4^+ present/absent. Reasons:

Steps 1-5

Ag^+ present/absent. Reasons:

Hg_2^{2+} present/absent. Reasons:

Pb^{2+} present/absent. Reasons:

Steps 6-12.

Pb^{2+} present/absent. Reasons:

Hg^{2+} present/absent. Reasons:

Bi^{3+} present/absent. Reasons:

Cu^{2+} present/absent. Reasons:

Steps 13-19.

Name_____   Section_____

Station_____   Date_____

General Unknown

Fe^{2+} or Fe^{3+} present/absent. Reasons:

Mn^{2+} present/absent. Reasons:

Cr^{3+} present/absent. Reasons:

Zn^{2+} present/absent. Reasons:

Steps 21-25.

Na^+ present/absent. Reasons:

Ba^{2+} present/absent. Reasons:

Ca^{2+} present/absent. Reasons:

Other observations.

Circle all ions you report as present

NH_4^+	Ag^+	Hg_2^{2+}	Pb^{2+}	Hg^{2+}
Bi^{3+}	Cu^{2+}	Fe^{2+} or Fe^{3+}		Mn^{2+}
Cr^{3+}	Zn^{2+}	Na^+	Ba^{2+}	Ca^{2+}

Experiment 44
Unknown Salt

Object

In today's experiment, you will determine the formula of a pure salt by dissolving the salt and performing qualitative analysis on the solution.

Background

You will be issued a solid salt of unknown composition. The salt is a pure compound containing one of the 14 cations you have studied and one of the 7 anions you have studied. You are to determine the ions which it contains and to suggest a possible formula for the salt.

The determination is done in a simple manner: dissolve the salt in water and perform qualitative analysis on the solution which results. If the salt dissolves in water, you should obtain positive results for only one of the cations and for only one anion. You can be assured that no "double salts" such as $Na_3[CrCl_3Br_3]$, $(CrCl_3 \cdot 3NaBr)$, will be used as unknowns.

The experiment may be more complex if certain salts are encountered. What should you do if the salt won't dissolve in water? In order to perform the tests, you must dissolve the salt to give a solution. If the salt won't dissolve in cold water, try heating the water. (What if the unknown salt were $PbCl_2$?) If the salt won't dissolve in hot water, centrifuge and discard the solution. Try to dissolve the unknown salt in dilute nitric acid. (What if the unknown salt were $CaCO_3$?) Perhaps some other reagent may be required to dissolve the salt. (What reagent would be required to dissolve AgCl?) When the unknown salt is placed in solution, you must remember how you made the salt dissolve, and what ions the dissolution process might have placed in the solution. Solutions obtained by dissolving a solid in nitric acid will always show a positive nitrate test, and yet you can see that it is possible that the unknown salt did not contain nitrate.

A simple solubility rule may be of value to you at this time. The rule states that: almost all ammonium, sodium and nitrate salts are soluble in water. If you have a salt which is insoluble in water, you can be reasonably sure that it is *not* a NH_4^+, Na^+ or a NO_3^- salt. You can add NaOH, NH_4OH or HNO_3 to cause the substance to dissolve and not encounter any problems. There are 98 salts which are possible unknowns, e.g., 14 nitrates, 14 chlorides, etc.; (14 cations × 7 anions = 98 salts). Each one will behave different, but you should now have the understanding to determine which salt has been issued to you. You should find this experiment interesting and enjoyable.

Procedure

Dissolve half of your solid, unknown salt to form a homogeneous solution. If the salt won't dissolve in cold water, try hot water, dilute HNO_3, dilute NH_4OH, aqua regia or some other reagent. When the salt has been dissolved, perform qualitative analysis on the solution to determine what anion and what cation are present. Suggest a possible formula for the unknown salt.

263

Remember that your unknown contains only one anion and one cation among those you have studied. No mixed salts are used as unknowns.

It is suggested that you perform anion analysis on separate portions of the solution. Then perform the ammonium test on a portion of the solution. Perform the cation analysis, if necessary, on 3 ml of the solution.

Name_____   Section _____

Station_____   Date_____

Unknown Salt

Report on Unknown Salt Number_____ .

Procedure to put salt in solution.

Anion Tests:

Sulfate

Carbonate

Phosphate

Chloride

Bromide

Iodide

Nitrate

Anion in salt is (circle one).

SO_4^{2-} CO_3^{2-} PO_4^{3-} Cl^- Br^- I^- NO_3^-

265

Ammonium Test:

Which Group does Cation belong to? Explain why you conclude this.

Which cation is present in the unknown? (circle one)

Ag^+ $\quad\quad$ Hg_2^{2+} $\quad\quad$ Pb^{2+} $\quad\quad$ Hg^{2+} $\quad\quad$ Bi^{3+} $\quad\quad$ Cu^{2+} $\quad$ Fe^{2+} or Fe^{3+}

Zn^{2+} $\quad\quad$ Cr^{3+} $\quad\quad$ Mn^{2+} $\quad\quad$ Ba^{2+} $\quad\quad$ Ca^{2+} $\quad\quad$ Na^+ $\quad\quad$ NH_4^+

What observations support this conclusion?

Cation_____

Anion_____ $\quad\quad$ Possible formula for salt_____ .